SAVE THE BAY'S
UNCOMMON GUIDE

to Common Life of Narragansett Bay
and Rhode Island Coastal Waters

Printed by Sheahan Printing Corporation.

ISBN-10: 0615229018
ISBN-13: 9780615229010

First edition published in 1998 as
"The Uncommon Guide to Common Life
of Narragansett Bay."

100 Save The Bay Drive, Providence, RI 02905   401-272-3540   SAVEBAY.ORG

SAVE THE BAY'S UNCOMMON GUIDE

to Common Life of Narragansett Bay
and Rhode Island Coastal Waters

2nd Edition

Editor
Donna M. DeForbes

SAVEBAY.ORG

Editor
Donna M. DeForbes

Contributing Writers
Jean Bambara
Marci L. Cole Ekberg
Wenley Ferguson
Rob Hudson
John Martin
David Prescott
John Torgan
Abby Jane M. Wood

Designer
Donna M. DeForbes

Cover Photo
Rob Hancock

Inside Photography
Margaret Bellucci
Donna M. DeForbes
Tom Freeman
John Martin
David Prescott

Project Contributors
Elizabeth Albert
Corinne Dedini
Suzanne Faubl
Jessica Harris
Pazzaz Hill
Rick McKinney
Leslie Munson
Matt Ortoleva
Jamie Shave
Carol Sullivan
Rebecca Travis
Carol Lynn Trocki

Illustrators
All illustrations by Riley Young, except for illustrations by:
Linda Cuddigan, pgs. 4(#9), 6, 10, 12, 14, 20, 22, 42
Nancy Chandler Edwards, pgs. 66, 132, 136, 140, 234, 262, 268, 270, 280
Gail Sirpinski, pgs. 288, 290
Jessica Waterman, pg. 106
The Connecticut Arboretum, pgs. 26, 28
U.S. Fish and Wildlife Service's *Fishes of the Gulf of Maine,* pgs. 154, 158, 172, 184, 186, 188, 208, 214, 216, 218, 220

Acknowledgements
Save The Bay's Uncommon Guide to Common Life of Narragansett Bay and Rhode Island Coastal Waters is a labor of love and a team effort. Many thanks to those who contributed to the research, development, scientific and technical review of the original edition:

Edwin Barber
Christy Law Blanchard
Nicole Cromwell
Betsy Dickinson
Anne DiMonti
Wenley Ferguson
Eve Formisano
Suzanne French
Lisa Gould
Mark Gould
Andrew Lipsky
Leni Machinton
Frederick D. Massie

Evan Matthews
Jeffrey Nield
Patty O'Biso
Adrienne O'Connor
Marisa J. Poli
Candace Powell
Christopher Powell
Marvin Ronning
Prentice Stout
John Torgan
Meredith Wall
Ellicot Wright
Riley Young

Table of Contents

Dedication

This book is dedicated to everyone who uses and enjoys Narragansett Bay – an integral part of Rhode Island's heritage. Save The Bay wrote this book to educate the public and to inspire a community of care around the Bay. It is the responsibility of all of us to preserve Narragansett Bay and its watershed for future generations.

Introduction to the Guide

Narragansett Bay is an estuary. Estuaries — where fresh and salt water meet — are among the most diverse and biologically productive ecosystems on earth.

Our Bay is a spawning ground, nursery, habitat, work place and playground for thousands of species of life, including winter flounder, lobster, hard-shell clams, eelgrass, seals and people. Yet many of those who live and work within the Narragansett Bay watershed are unfamiliar with its wildlife and know little about the interrelationships between the Bay ecosystem and human actions.

Save The Bay's Uncommon Guide to Common Life of Narragansett Bay and Rhode Island Coastal Waters serves as a bridge between the people who use and enjoy Narragansett Bay and some of the more common plants, invertebrates, fish, birds and marine mammals that share our Bay. Although much of the information contained in this Guide was drawn from a range of scientific texts and papers, it was written to be understood and appreciated by individuals without extensive scientific backgrounds.

This 2nd edition reflects the expanded scope of Save The Bay's work — we have moved from a focus on Narragansett Bay itself to the whole region, including the unique and intimately connected freshwater tributaries and coastal marine waters.

The *Uncommon Guide* is divided into five major sections: Plants, Invertebrates, Fish, Birds and Mammals, with each section further subdivided into several general categories. Every entry includes an illustration, species description, field markings and quick-reference icon on where you might find a particular species. An expanded explanation of these entry descriptions can be found on the next page.

Entries were selected through a review process and reflect our consensus of some of the most common forms of Narragansett Bay life. For those interested in learning more about these species or about the thousands of other animals and plants found in and around the Bay, we recommend a review of the books included in the *Uncommon Guide*'s bibliography as an excellent starting point.

An educational tool for adults and children to use alone or together, we hope our *Uncommon Guide* will inspire interest in Narragansett Bay and encourage a greater understanding of and appreciation for Bay protection and restoration.

ENTRY DESCRIPTION

Names

A species' scientific name can be found here as well as other nicknames commonly used in reference to a plant or animal.

Illustration

The black-and-white illustrations are aids to visual identification. Illustrations are not drawn to scale, but instead highlight as much detail as possible for each species. They are complemented by descriptions of other physical identifications, including color, size and distinguishing features. Some illustrations include pictures of distinctive egg cases, as in the entries for skates, whelks and moon snails, while other illustrations depict a variety of related species, as in the entries for zooplankton and phytoplankton.

Distinguishing Features and Behaviors

Information contained in this section details specific physical characteristics and is designed to complement the illustrations. Behaviors noted are characteristic of the animal and are included to help with identification. Predation, mating and defensive behaviors are mentioned, where appropriate.

Relationship to People

This section covers connections between plant or animal populations and historic or current human actions. It also indicates plant and animal sensitivity to human action. Where appropriate, commercial and recreational values are highlighted. Particular attention is paid to specific connections to the Narragansett Bay region and to the environmental impacts of climate change and human activity, such as water pollution, habitat alteration and overfishing.

Field Markings

This section mentions a species' size, prominent color and markings and, sometimes, visual variations between gender. Unless otherwise noted, descriptions refer to mature animals or plants.

Habitat

For the purpose of the *Uncommon Guide*, we have created a quick-reference icon depicting the area where one might find a particular species. The icon reflects a species' *predominant* habitat; other habitats are noted within the text. Icons depict one of three common Narragansett Bay habitats: ocean, estuary and shoreline (includes salt marshes, beaches and tide pools).

Seasonal Appearance

This section reflects the period of time when a species is commonly found in Narragansett Bay.

Sensitivity Level

This gauge indicates the degree to which a species responds to human influence. Sensitivity includes both positive *and* negative responses. Levels were determined based on a subjective review of factors, including pollution, habitat alteration and overfishing. A range of shaded circles (as depicted below) is used to denote sensitivity levels as either low, medium or high. An exclamation point within the high-level circle depicts severe sensitivity, as is the case with the American eel and the Atlantic salmon.

About Save The Bay

*The Save The Bay
Center in Providence
is open to the public
weekdays, 8:30 - 4:30.
Our grounds are open
from dawn until dusk.
For more information
on our work in the
Bay Community, call
401-272-3540 or visit
SAVEBAY.ORG.*

Save The Bay began in 1970 when a group of concerned citizens recognized that the best way to protect Narragansett Bay was to unite and fight to keep the Bay clean and healthy. Since then, Save The Bay has grown into one of the largest and most effective non-profit, membership-based environmental organizations in New England.

We operate from three waterfront locations across Rhode Island: the award-winning green-designed Save The Bay Center in Providence, our aquarium and marine science Exploration Center in Newport and our South County Coast office in Westerly.

Save The Bay is committed to fostering a personal connection between people and Narragansett Bay and encouraging investment in the Bay's future. Save The Bay and its supporters are an active force in southern New England, working to protect, restore and explore Narragansett Bay and its watershed. Together, we:

- Advocate for better Bay management on the local, state and federal levels.

- Restore Bay habitat — salt marshes, eelgrass beds, fish runs and shorelines — to healthier and functioning conditions.

- Encourage discovery and knowledge of the Bay through our educational programs.

- Seek to create a community of Bay citizens who will preserve the Bay for generations to come. One way we are trying to reach this goal is by enabling a communications and action network at SAVEBAY.ORG.

Due to Save The Bay's efforts — coupled with the work of other individuals, groups, businesses and government agencies — most major sources of Bay pollution have been dramatically reduced over the past 38 years.

When *Uncommon Guide* was first published, we wrote that stormwater runoff, failing septic systems and poorly managed development were taking their toll on the Bay, and that many areas of Narragansett Bay were closed to swimming and fishing. While this is still true, Save The Bay has played a major role in addressing these problems. We advocated strongly for the Combined Sewer Overflow project that, when it goes online in 2008, will greatly reduce the stormwater runoff problem in the upper Bay. In 2007, we fought successfully for passage of cesspool phase-out legislation that will begin a process of eliminating cesspools and replacing failed or outmoded septic systems. In 2004, Save The Bay lobbied for new nitrogen discharge permits at the Fields Point and Bucklin Point wastewater treatment facilities. The limits we fought for will help reduce the number and intensity of clam and fish kills in the upper Bay as well as reduce the abundance of algae and sea lettuce that chokes the shoreline in warm weather.

Much has changed since the first edition of the *Uncommon Guide* came out in 1998.

Save The Bay's Exploration Center and aquarium, located on Newport's Easton's Beach, is home to over 150 local species. The interactive, hands-on Center is open to the public daily during the summer. Call 401-324-6020 for school and group appointments year-round.

The Save The Bay Center at Fields Point, which opened in 2005, has become the hub of the Bay Community. In addition to our busy classrooms, the space is used by a cross section of groups working to protect the Bay and its watershed. We have expanded our summer camps and added new public programs that help connect people with the Bay.

Weekend tours are now available, allowing visitors to see the green building techniques we have employed as well as the work we have done to restore a shoreline that was previously a municipal dump. Our green roof is part of our extensive stormwater management system and its 12,000 plants help reduce stormwater runoff. Our coastal buffer includes 20 species of drought-tolerant and salt spray-tolerant native trees and shrubs that protect against shoreline erosion. The buffer has become a habitat for creatures from snow buntings and canada geese to wild turkeys and spotted sandpipers. In short, we have transformed a landfill into a landmark for responsible shoreline development.

In 2006, we opened the Save The Bay Exploration Center in Newport. The aquarium and interactive learning center — originally operated by the New England Aquarium — now features more than 150 species of Bay creatures. Located on Easton's Beach, the facility is within walking distance of nearby salt marshes. Also in Newport, our winter seal watch tours remain one of the region's most popular seasonal attractions.

Through our Baykeeper program, we continue to work in the public and private sectors to affect positive change for the Bay. In 2007 we expanded our reach to Little Narragansett Bay and the Coastal Ponds region by launching our South County Coastkeeper program and office in Westerly.

Institutionally, there has been a significant change. For many years, our mission of advocacy, habitat restoration and education was expressed as "Protect, Restore and Explore." In 2007, we added an important fourth verb to the phrase — Save The Bay now works hard to "Connect" people to Narragansett Bay. Whether it's a seal watch tour, a sunset cruise from the Bay Center, a shoreline cleanup, a habitat project or a State House rally, we're working to become the catalyst that connects people to the great natural resource we all love.

And so, in its own way, our *Uncommon Guide* helps people of all ages connect with the Bay through a better understanding of its fascinating creatures and their habitats. This book is what Save The Bay is all about.

Narragansett Bay — Why Save Me?

The land surrounding Narragansett Bay and its three major rivers – the Pawtuxet, the Blackstone and the Taunton – is known as the Narragansett Bay watershed and encompasses 1,853 square miles. Although the Bay is located primarily in Rhode Island, 61% of the watershed exists in Massachusetts.

I am a designated estuary of unparalleled beauty and national significance. Estuaries — where fresh water from the rivers meets salt water from the sea — are among the most diverse and biologically productive ecosystems on earth, surpassing tropical rainforests. Yet I have become the final resting place for sewage, oil, chemicals, fertilizer and other pollutants. *Your contribution helps Save The Bay to oppose destructive coastal development projects and to advocate for better laws and regulations toward keeping me clean.*

I am spawning ground, nursery and habitat for more than 60 species of fish and shellfish, more than 200 bird species, plus marine mammals such as seals, dolphins and sea turtles. Due to human pollution, some of my most important marine species have declined dramatically, including winter flounder, scallops, lobster, cod and tautog. Where more than 1,000 acres of eelgrass once covered my floors, today, less than 290 remain. *Your contribution helps Save The Bay to restore productive eelgrass beds, which are the nurseries and spawning grounds for threatened fish and marine life.*

I provide recreational opportunities to millions of people in Rhode Island and Massachusetts who live, work and play within my watershed. I welcome more than 100,000 fishermen each year, and over 32,000 recreational boats cruise my waters. However, my beaches close frequently due to high levels of pollution. Every summer, I take in too much nitrogen from wastewater and polluted runoff; it robs me of oxygen, leading to fish kills and foul odors. *Your contribution helps Save The Bay to fight for tougher limits on pollution from wastewater treatment plants so you can continue to enjoy my waters.*

I offer wetlands and marshes that help shield my shorelines from storms and act as the most effective defense against flooding. They also serve as giant pollution sponges that prevent harmful contaminants from entering drinking water supplies. Approximately 50% of my salt marshes are gone and the rest are degraded, thereby eliminating thousands of acres of critical wildlife habitat. *Your contribution helps Save The Bay to identify, plan and fund the restoration of degraded salt marshes so that property is protected and drinking water is safer.*

I am a living classroom and my future is in the hands of the next generation of Bay users and policymakers. Thousands of kids meet me through Save The Bay's classroom and shipboard programming where they discover endless possibilities for exploration. *Your contribution helps Save The Bay to continue educating your children and grandchildren so that all future generations can enjoy a well-cared-for, healthier Bay.*

I provide a quality of life that attracts businesses, industries and more than 12 million visitors each year, generating thousands of jobs and billions of dollars for the regional economy. For commercial fishermen, I am a way of life, producing about eight million pounds of quahogs annually, with a value of $6 million. Unfortunately, over 40% of my waters are closed to shellfishing either permanently or on a conditional basis because of bacterial pollution. Fish can reach their spawning grounds in only 21 of the 45 historic fish runs in my watershed. *Your contribution helps Save The Bay to restore habitat by building fish ladders that allow fish to reach their native rivers, which helps to keep our marine economy strong.*

I am an environmental treasure. I connect all of Rhode Island's and southeastern Massachusetts' communities together and to the ocean. I am the staple of the area's tourism industry, the greatest public asset and the symbol of the region's history and culture.

℥

Formed by melting glaciers over 5,000 years ago, Narragansett Bay is connected to Rhode Island Sound and the Atlantic Ocean by three drowned river valleys: the East Passage, the West Passage and the Sakonnet River. Twenty-five miles long, ten miles wide, with a surface area of 147 square miles, the Bay holds 706 billion gallons of water at mid-tide.

With an average depth of 26 feet, Narragansett Bay is relatively shallow. The tidal range of the Bay is three to four feet every 12.5 hours, with the tide taking about 20 minutes to move from the mouth of the Bay to the Bay's upper reaches. Through the salt water tidal action from the Atlantic Ocean and the daily inflow of 2.4 billion gallons of fresh water a day from rivers and rainfall, the Bay flushes, on average, every 26 days.

In the 1970s, a swimmable and fishable Narragansett Bay was laughable. With the help of our members and volunteers, Save The Bay has made tremendous progress in working for the Bay people want. Conventional pollution from wastewater treatment plants has been reduced by more than 60% since the 1980s, and toxins in wastewater have been reduced by more than 90%, resulting in the return of marine and wildlife as well as clearer, better-smelling water for some. But we still face daunting challenges. Pollution from wastewater, stormwater and power plants continue to put these gains in jeopardy.

Anchored in the belief that people value what they understand and appreciate, Save The Bay's *Uncommon Guide* is designed to promote a better understanding of Bay life and our role in its ecosystem. It aims to foster a personal connection between people and Narragansett Bay.

HOW CLIMATE CHANGE AFFECTS NARRAGANSETT BAY

Since this guide was last published, climate change has been recognized as an important factor in coastal ecosystems — its effects locally are already apparent.

Climate change, also known as global warming, is the warming of the atmosphere due to an increase of greenhouse gases. It is a phenomenon that affects us in a myriad of ways, including droughts, floods, increases in infectious disease vectors (such as mosquitos) and increases in hurricane intensity. Warmer water temperatures cause ice caps to melt and oceans to expand, resulting in sea level rise.

In Narragansett Bay, the water temperature has increased 3° F since 1900 (Oviatt 2004). In that same time, Bay waters have risen up to eleven inches. The rate of rise has more than doubled from 1 millimeter per year in the mid-1800s to a current rate of 2.6 millimeters per year (Donnelly and Bertness 2001). While these numbers may sound small, the changes in water temperature and sea level have dramatic impacts to the animals and plants of Narragansett Bay.

With this in mind, we have updated the *Uncommon Guide* to reflect current knowledge about how climate change is already affecting some local species.

OCEAN

 ESTUARY

 SHORELINE

WHERE DO I FIND IT?
Use these icons as a quick reference to where you might find a particular Bay species. Icons denote a species' predominant habitat; other preferred habitats are mentioned within the text.

Plants

Plants provide the basis of life in Narragansett Bay. From phytoplankton and seaweeds to seagrasses and coastal marsh species, they provide oxygen, food, and habitat for many marine animals, including birds, fish and invertebrates.

Marine plant species are diverse, colorful and hearty, ranging in size from microscopic plankton to giant kelp. The majority of plant life is composed of algae and seaweeds. Pigments present in algae are responsible for their variety of color: blue-green, green, red and brown.

Although not as complex as land plants, algae still photosynthesize. They are important producers of organic material and serve as the primary food source for many different animals in the Bay.

Seaweeds, a type of algae, do not have roots; instead they have structures called holdfasts, which allow them to attach to sturdy surfaces like rocks and shellfish. Rather than taking in water from roots, seaweeds absorb nutrients, carbon dioxide and water through their surfaces. Like land plants, seaweeds may be annual, perennial or year-round. They tend to change color when dead or washed ashore, therefore color is not always the best identifying characteristic. Here, we mention brown, green and red seaweeds, highlighting some of the most common ones in Narragansett Bay.

Seed-marine plants have developed roots, flowers and seeds. Eelgrass, unlike algae and seaweeds, is a submerged seed-bearing plant complete with roots and flowers.

Coastal and shoreline plants are also seed-bearing, flowering plants. Marsh grasses, pickleweed and reeds are some typical examples. Marsh grasses can tolerate a saline environment and have mechanisms either to exclude salt from entering their systems or to excrete salt taken out of the water.

Eelgrass

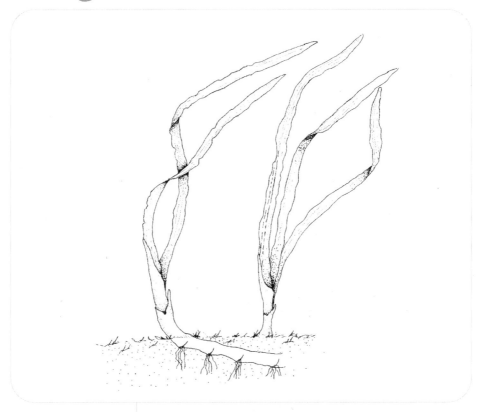

COLLOQUIAL NICKNAMES:
Seagrass, Submerged
Aquatic Vegetation (SAV)

SCIENTIFIC NAME:
Zostera marina

Distinguishing Features and Behaviors

Eelgrass is a flowering underwater plant with quarter-inch wide leaves that can reach lengths of three feet. Eelgrass grows in distinctive clumps, known as beds, in near-shore waters at depths ranging from four to nine feet. Eelgrass beds are always completely submerged, and their roots (or rhizomes) anchor the grass to sandy or muddy bottoms. From spring through late summer, eelgrass produces hundreds of seeds, which float with the current before sinking to the bottom.

A true flowering plant commonly submerged at high tide, eelgrass is sometimes misidentified as saltmarsh cordgrass, a plant that grows at the edge of the shore in the intertidal zone.

Uprooted eelgrass can sometimes be found in bright green, tangled clumps or individual strands along the shore. Dead eelgrass appears brown or almost black and, when left out of water for a period of time, it becomes dry and papery.

Eelgrass is one of the Bay's most vital habitats for a wide range of fish and wildlife, including flounder, scallops and crabs. At the base of the food chain, many species of commercially valuable fish feed on or take shelter in these beds at some point in their lives. Eelgrass beds filter excess nutrients out of the water and help prevent shoreline flooding and erosion by stabilizing sediment and buffering wave action. Because it requires specific amounts of light and clean water, the presence of eelgrass is an indicator of healthy water quality.

Relationship to People

Historically, eelgrass beds flourished in many areas of Narragansett Bay and helped support a thriving commercial scallop industry. Increased water pollution, shoreline development, boat traffic, eelgrass wasting disease and hurricane damage have significantly reduced the Bay's eelgrass beds. This loss has affected Bay fish and wildlife populations and has virtually eliminated commercial scalloping in the Bay. Eelgrass is also sensitive to increased water temperature due to climate change.

In 2006 and 2007, aerial photographs and ground-truthing were used to map existing eelgrass beds in Narragansett Bay and Block Island Sound. The information gathered will be used to develop programs to restore and expand eelgrass beds through transplant projects and improved environmental management.

FIELD MARKINGS: Grows in clumps; can form extensive underwater beds or meadows. The green leaves are quarter-inch wide. Size: up to 3 feet long.

HABITAT: Found in brackish to entirely salt water of protected inlets and bays. Grows in sandy, silty or gravelly substrates in the subtidal zone.

SEASONAL APPEARANCE: Blooms early spring and summer.

SENSITIVITY LEVEL:

Phytoplankton

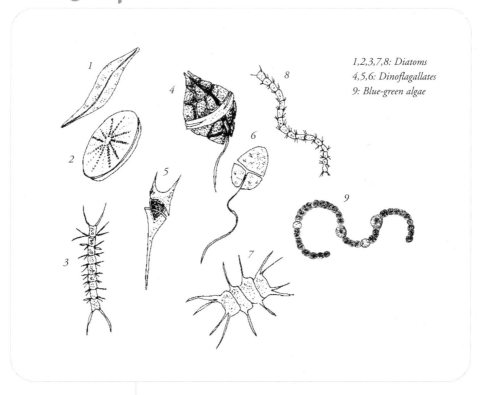

1,2,3,7,8: *Diatoms*
4,5,6: *Dinoflagallates*
9: *Blue-green algae*

Distinguishing Features and Behaviors

Phytoplankton are the foundation of life in Narragansett Bay, serving both directly and indirectly as food sources for all living animals. They can be single-celled or colonial, free-floating plants. Some phytoplankton are non-swimming; others have various forms of locomotion, from the ability to migrate vertically in the water column to having extended projections that increase surface area and slow down the rate at which they sink.

Zooplankton — small animals and larval forms of fish and invertebrates — together with phytoplankton make up the group called plankton.

The predominant forms of phytoplankton are: diatoms, golden-brown algae, green algae, blue-green algae and dinoflagellates. Over 10,000 species of diatoms alone exist in the world. They have an exoskeleton made of silica and have no means of swimming; their movement through the water depends solely on currents.

Blue-green algae are a dominant freshwater phytoplankton, while diatoms make up the majority of phytoplankton in water of higher salinity. Phytoplankton use the process of photosynthesis to sustain themselves, converting carbon dioxide into oxygen and organic materials. They are believed to produce 80% of the organic material in the world. Phytoplankton are consumed by a wide range of fish and invertebrates in the Bay.

Relationship to People

Phytoplankton form the basis of life in the world's oceans. By converting sunlight and inorganic elements into plant tissue, these tiny cells begin the complex food web that ultimately sustains all fish, invertebrates, mammals and birds. Phytoplankton determine the color, odor and taste of saltwater and are a fundamental measure of productivity in the marine environment.

FIELD MARKINGS:
Most individual structures are microscopic.

HABITAT:
Found in abundance in open water, estuaries and along the shoreline.

SEASONAL APPEARANCE:
Year-round with blooms in the spring and fall.

SENSITIVITY LEVEL:

Seaweeds, Brown

Spiny Sour Weed

COLLOQUIAL NICKNAMES:
Spiny Sour Weed,
Fingered Kelp, Edible Kelp,
Ralfsia, Gulfweed

SCIENTIFIC NAME:
Desmarestia aculeata,
Laminaria digitata,
Alaria esculenta,
Ralfsia verrucosa,
Sargassum filipendula

Distinguishing Features and Behaviors

Mostly marine and benthic, many brown seaweeds exist in an upright form, like kelp, or in a crust form, like *Ralfsia*. They range in color from near-black to yellow-brown to dark green. Some species of kelp can grow to over 160 feet long, making the brown seaweeds the largest algae on earth.

Many of these species are found attached to rocks and other sturdy substrate in the intertidal to the subtidal zone. Brown seaweeds are used by marine organisms as a source of food or as habitat.

A common form of kelp found in Narragansett Bay is fingered kelp *(Laminaria digitata)*, which has fingerlike blades extending from the short, thick stipe. Edible kelp *(Alaria esculenta)* — a long, thick seaweed, brown to yellow-green in color — is usually found along Cape Cod and northward. Edible kelp grows three to five feet long with a spine, or midrib, running along the length of its ribbonlike blades. On the West Coast of the United States, a unique ecosystem of large kelp forests can be found.

Spiny sour weed *(Desmarestia aculeata)* is a brown seaweed that comes in two forms, depending on the season. In spring, it is light brown with fine filaments attached to the branching; in summer, its spiny branchlets have an alternating pattern, and dark brown stems give way to lighter brown branches.

Ralfsia verrucosa is a dark brown to orange crust that forms on rocks. Circular and smooth in shape when young, it becomes brittle, irregular and rough as it ages.

Gulfweed *(Sargassum filipendula)* ranges in color from golden-brown to dark brown. This seaweed has narrow blades with serrated edges, a midrib and small dark spots. Small, round pealike air bladders may also be present.

Relationship to People

Brown seaweeds are harvested worldwide by the tons. Alginic acid is extracted from the harvested seaweed and used in the production of toothpaste, ice cream, soaps, smoothing paper and conserving dyes for fabric. Some kelps are eaten by humans or used as animal feed because they are a good source of protein, carbohydrates, fibers and other nutrients.

Rockweed and Knotted Wrack

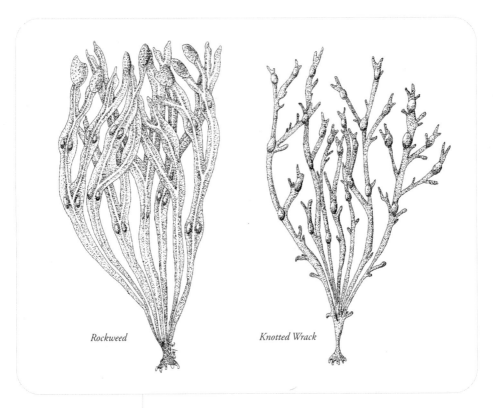

Rockweed   Knotted Wrack

COLLOQUIAL NICKNAMES:
Bladder Wrack

SCIENTIFIC NAME:
Rockweed:
Fucus vesiculosus
Knotted Wrack:
Ascophyllum nodosum

Distinguishing Features and Behaviors

Rockweeds and knotted wrack are large brown seaweeds that can grow up to three feet long. They feel and look like leather. The unique feature of both rockweed and knotted wrack is the presence of air bladders along the length of their blades. The air bladders give these seaweeds needed buoyancy in the water column to help the plants absorb sunlight.

Bladder rockweed, or bladder wrack *(Fucus vesiculosus)*, has equal-sized, branching, straplike stems lined with well-developed, paired air bladders. It has midribs, which are spinelike structures along the length of the blades. The flat, yellowish branch tips are the youngest part of the rockweed.

Knotted wrack gets its name from the irregularly spaced air bladders resembling knots along its branches. Unlike bladder rockweed, knotted wrack has no midrib.

All species of rockweeds and knotted wrack are true seaweeds and do not have roots. They are found attached to rocks, shells and artificial structures via a holdfast, or rootlike structure. As long as the holdfast remains intact, rockweed will regenerate even if it is torn off the rocks. Rockweed and knotted wrack grow in dense masses on rocks and other formations. They obtain nutrients from the surrounding seawater through pores in their blades. In the spring, rockweed and knotted wrack reproduce by growing short, yellow or orange fruiting structures that release sex cells and then fall off the plant.

Relationship to People

Rockweed and knotted wrack serve as excellent pollution monitors because they are sessile, or nonmoving, and concentrate heavy metals from the water in their tissue. As a result, these seaweeds are sometimes used to measure long-term pollution levels.

They are used commercially as poultry meal, fertilizer and garden mulch. A compound called alginate is extracted from their dried form and used as a thickener in paints, foods and cosmetics.

Rockweed and knotted wrack often wash up in large clumps and can be dangerous to beach strollers. The algae is slippery and can cover up hazardous, rocks and crevices that would normally be avoided.

FIELD MARKINGS:
Both species are brown to olive green in color with air bladders along the branches. Bladder rockweed has paired air bladders and a midrib. Knotted wrack is ribless with irregularly spaced air bladders. Size: up to 3 feet long.

HABITAT:
Open water, shallow brackish water, coves and tide pools; attached to rocks.

SEASONAL APPEARANCE:
Year-round.

SENSITIVITY LEVEL:

Sugar Kelp

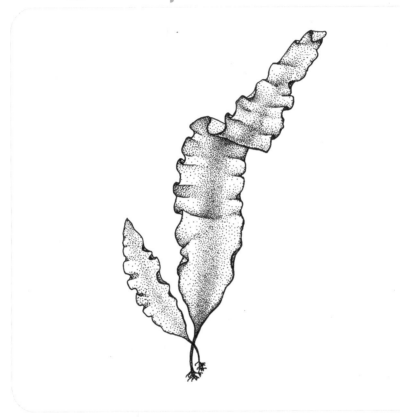

COLLOQUIAL NICKNAMES:
Kelp

SCIENTIFIC NAME:
Laminaria saccharina

Distinguishing Features and Behaviors

Kelp is a brown seaweed, usually brown to yellow-green in color. There are many species of kelp, including the giant kelp beds of the Pacific, but sugar kelp is the species common to Narragansett Bay.

Resembling lasagna noodles, the blades are ribbon-like, usually several feet long, but can grow up to ten feet. Kelp does not have true roots anchored in the sediment; instead it attaches itself to rocks and shells by a holdfast — a strong rootlike structure that serves as an anchor. The holdfast keeps kelp from being washed ashore during heavy wave action, although in very strong storms, the plant will often break at the stem, washing ashore and leaving the holdfast still attached to the rock.

Some types of kelp have a hollow midrib or tiny air bladders in the blades that allow the plant to float up in the water column. Kelp obtains nutrients from the water through pores in its blades, unlike terrestrial plants that get nutrients through a root system.

Kelp is primarily a subtidal plant but is often found on rocks in the intertidal zone during low tide. Sugar kelp and edible kelp *(Alaria esculenta)* can be found along the New England coast.

Another common species of kelp in Narragansett Bay is fingered kelp *(Laminaria digitata)*, distinguishable by many finger-like blades extending from the holdfast.

Relationship to People

Sugar kelp is related to the kelps of the Pacific Ocean that can reach lengths of over 100 feet. Both forms are used as soil fertilizer.

The "sacchar" in *saccharina* means sugar. Because they are particularly rich in vitamin C, kelp blades can be ground up and used as salt or vitamin supplements. The midrib of kelp is often used in salads or made into candy. A high demand for kelp has caused restrictions on harvesting to protect the species.

FIELD MARKINGS:
A long, thick seaweed, brown to yellow-green in color; blades may be ruffled along the edges. Size: averages 3 to 6 feet long.

HABITAT:
Exposed shores, open water, tide pools; attached to rocks. Commonly found in clumps along the high-tide line of the coast.

SEASONAL APPEARANCE:
Year-round.

SENSITIVITY LEVEL:

Seaweeds, Green

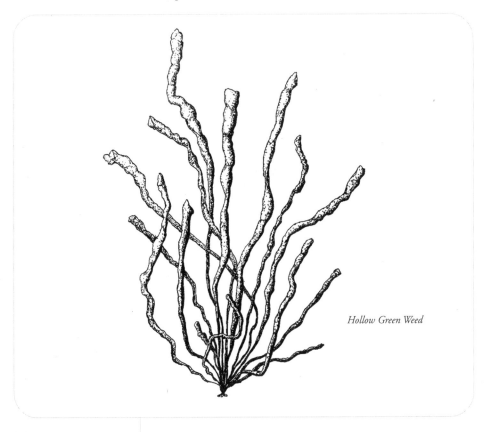

Hollow Green Weed

COLLOQUIAL NICKNAMES:
Spaghetti Algae,
Hollow Green Weed

SCIENTIFIC NAME:
Chaetomorpha linum,
Ulva intestinalis

Distinguishing Features and Behaviors

The group of green seaweeds consists of many diverse species. Although most predominant in fresh water, green seaweeds have many marine representatives, including green fleece, hollow green weed, spaghetti algae and sea lettuce.

Spaghetti algae *(Chaetomorpha linum)* is a light to dark green coarse algae that is filamentous and unbranched, with a single row of cells. It is about four inches long, and the cells are visible with a magnifying glass.

Spaghetti algae is found on sandy or muddy bottoms, from the intertidal to the subtidal zone. It is usually entangled and looks like discarded fishing line.

Hollow green weed *(Ulva intestinalis)* is pale green or yellow in color, hollow and unbranched. The branches are about one-quarter inch to one inch in width and up to ten inches long. Sometimes called by its old Latin name, *Enteromorpha intestinalis*, hollow green weed is named for air bladders in its branches, which enable the seaweed to float upright in the water.

Commonly found attached to rocks, shells and other algae, hollow green weed prefers brackish to very salty water. As a result, it is tolerant to changes in salinity and can occasionally be found in water that is nearly fresh.

Relationship to People

The *Chaetomorpha* species are used in salt water aquariums as a way to filter out excess nitrates and other nutrients.

Hollow green weed thrives in waters that are moderately polluted; the weed can even tolerate living in severely polluted waters. It is used as a food source in some Asian and European countries.

Dead Man's Fingers

COLLOQUIAL NICKNAMES:
Green Fleece, Oyster Thief

SCIENTIFIC NAME:
Codium fragile

Distinguishing Features and Behaviors

Dead man's fingers, also known as green fleece, is a tubular seaweed, dark green to yellow in color, with a thick, spongy texture. It can grow up to three feet long, with branches up to one half inch in diameter. The coarse, bushy branches grow in Y-shaped forks, resembling green fingers or a mass of worms.

When clustered, green fleece resembles a carpetlike mat, and it is often found washed ashore in ropy masses. When dried by the sun, it can become white or gray in appearance. This seaweed uses a holdfast to attach itself to a wide range of objects, from rocks to living shellfish.

Green fleece is coenocytic, which means that it is made of a large multinucleate cell without cross walls, except in reproductive areas of the seaweed. It is able to rejuvenate from broken fragments. During the winter, this seaweed stores excess nitrogen so it doesn't have to worry about this nutrient in the summer when in its reproductive phase.

Relationship to People

Green fleece was introduced from the Pacific Ocean to the North Atlantic around 1957, most likely attached to the bottom of European ships. Some of the seaweed's traits make it able to out-compete vital local habitats, such as eelgrass and kelp beds.

Dead man's fingers is often called "oyster thief." It will frequently attach itself to an oyster or scallop, and air bubbles — developed during photosynthesis — cause both the plant and mollusk to float above the bottom, which can eventually kill the shellfish. It is common to find a dead oyster or other mollusk washed ashore attached to green fleece.

FIELD MARKINGS:
Large, dark green seaweed, tubular and fingerlike; branching in Y-shaped forks. Grows in clumps. Size: up to 3 feet.

HABITAT:
Salt marshes, open water, tide pools and rocky shores; attached to rocks and shells.

SEASONAL APPEARANCE:
Year-round.

SENSITIVITY LEVEL:

Sea Lettuce

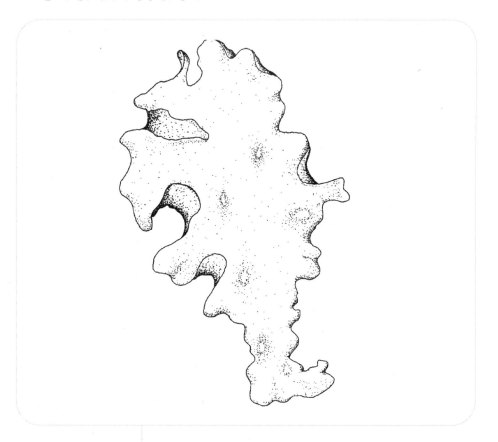

SCIENTIFIC NAME:

Ulva lactuca

Distinguishing Features and Behaviors

Sea lettuce is a bright green algae composed of lobed, ruffle-edged leaves that are coarse and sheetlike, resembling a leaf of lettuce. The leaves may appear flat, thin, broad and are often rounded or oval, typically perforated with holes of various sizes. Almost no stalk exists at the point of attachment, and no true roots are present. Sea lettuce may be found attached to rocks and shells by a holdfast, but it is usually free floating.

Among the most familiar of shallow water seaweeds, sea lettuce is usually found in areas of exposed rock and in stagnant tide pools. It has also been recorded at depths of 75 feet or more. Sea lettuce grows in both high and low intertidal zones and marshes throughout the year. When dried by the sun, its color can range from white to black.

Relationship to People

Tolerant of nutrient loadings that would suffocate many other aquatic plants, sea lettuce can actually thrive in moderate levels of nutrient pollution. Large volumes of sea lettuce often indicate high levels of pollution. Growth is also stimulated by the presence of other pollutants and is often found in areas where sewage runoff is heavy.

As a result, sea lettuce is used as an indicator species to monitor pollution trends. The density and location of this algae often indicates the presence of high amounts of nutrients. In areas where there is a high concentration or "bloom" of sea lettuce, sunlight is unable to reach other submerged vegetation (such as eelgrass), preventing photosynthesis and often killing the vegetation below. When sea lettuce dies, bacteria feeding on the decomposing sea lettuce use up a tremendous amount of oxygen in the water. This, in turn, depletes the oxygen available to other species, suffocating or driving them away.

Masses of sea lettuce can hamper swimmers and foul lines and fishing nets, but it does provide a home to some invertebrates, such as amphipods. Like lettuce grown on land, it can be used in salads and soups. Sea lettuce is also used to make ice cream and medicine.

FIELD MARKINGS:
Green seaweed,
sheetlike in appear-
ance. Can be white
or black when dry.
Size: ranges from
6 inches to 2 feet
in diameter.

HABITAT:
Estuaries, high and
low intertidal zones,
to depths of 75 feet.

SEASONAL APPEARANCE:
Year-round, with large
blooms in the summer.

SENSITIVITY LEVEL:

Seaweeds, Red

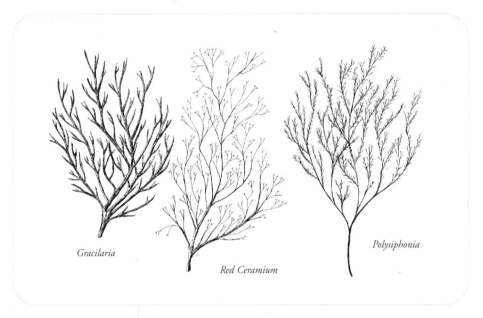

Gracilaria

Red Ceramium

Polysiphonia

Distinguishing Features and Behaviors

Red seaweeds make up the largest group of algae in the plant kingdom, with more species accounted for than brown and green seaweeds combined. They are almost exclusively marine plants. Although generally found in shallow waters (including tide pools, coves and eelgrass beds), red seaweeds are also able to withstand deep water and low-light conditions.

Red seaweeds can occur as large, branched plants or as bushy growths on rocks and shellfish. They range from five to sixteen inches long. There are many species of red seaweed in the North Atlantic and New England waters, but three species are commonly seen along the shores of Narragansett Bay and Rhode Island: *Polysiphonia*, *Ceramium* and *Gracilaria*.

Polysiphonia is the most common genus of red seaweed in the Rhodomelaceae family. The name is Latin for "many tubes," which aptly describes the structure of the branches. Often called "mermaid's hair," *Polysiphonia* is a bushy seaweed, growing in tufts that are yellow, pink, red or black. The top of the seaweed has many branches and feels soft to the touch compared to its coarse, dark bottom. It occurs mainly in protected shallow waters, often attaching itself with a holdfast to stones, shells and eelgrass. *Polysiphonia* is commonly found in Narragansett Bay as an epiphyte (one plant living on another plant) on knotted wrack.

Ceramium is a branched red seaweed, resembling *Polysiphonia*, but with pincerlike structures at the end of each of the branches. Its axes and branches have regular banding patterns. *Ceramium* will grow up to 16 inches long, often occurring in large tufts attached to just about any kind of substrate.

Gracilaria is a course red seaweed that can appear yellow-green or brownish purple. It has rounded or flattened, rubbery branches. *Gracilaria* is common in quiet, shallow waters and is most abundant in the warm summer months when it can be found floating over muddy or sandy Bay bottoms.

Another common genus of red seaweed in the Bay is *Porphyra*. Also known as laver, *Porphyra* may be seen as a very thin blade of reddish algae that can be narrow and long, short and broad, or roundish in shape. Edges of the blade may appear ruffled.

Relationship to People

Many species of red seaweed are commonly harvested for food. Usually eaten raw or dried, red seaweed is used in salads, soups and sushi. *Gracilaria* is harvested to make agar, a compound used in medical and biological research to culture bacteria and yeast. Agar is also used in the production of cosmetics and hand cream.

Coral Weed

Detail of branches

SCIENTIFIC NAME:
Corallina officinalis

Distinguishing Features and Behaviors

Coral weed maybe found as undergrowth beneath larger seaweeds or in tide pools and protected crevices. The featherlike branches of coral weed are very sensitive to drying out, so you will not find this species in areas where it would be out of the water for more than an hour.

Coral weed's branches are made of calcium carbonate, which gives the seaweed its rough texture. The plant ranges from pink to purple-red. Coral weed can build up sediments in its tufts and is used as a habitat for various small animals, which feed on the even smaller microorganisms that inhabit this algae.

Although it does not happen often, sea urchins have been known to graze on coral weed when some of their other food sources, such as kelp, are unavailable.

Relationship to People

Coral weed is one of the only calcareous species (having a rough, stonelike texture) found in Rhode Island waters. Similar species are more common in warmer waters, where broken pieces contribute to the lime-sand beach sediment.

This algae has been used in the medical field historically as vermifuges — a medicine that expels parasitic worms — and it currently aids in bone replacement therapy. Coral weed is also used in the cosmetics industry.

FIELD MARKINGS:
Pink to purple-red calcified segments with white tips. May appear as pure white beads when washed ashore. Size: up to 4 inches long.

HABITAT:
Attached to rocks in the intertidal zone.

SEASONAL APPEARANCE:
Year-round.

SENSITIVITY LEVEL:

Grateloupia

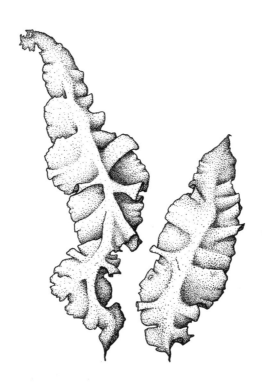

SCIENTIFIC NAME:
Grateloupia turuturu

Distinguishing Features and Behaviors

Grateloupia turuturu — formally known as *Grateloupia doryphora* — is an invasive species and a threat to local red algae like Irish moss and dulse *(Palmaria palmate)* because *Grateloupia* reproduces easily, crusts over in winter and, if blades detach from the substrate, they can reattach elsewhere, beginning the life cycle anew.

Grateloupia can also block out sunlight to other algal species in the intertidal zone. Local invertebrates such as snails feed on native algal species like Irish moss, so it is important to study *Grateloupia* to better understand the affect it has on Narragansett Bay and its inhabitants.

Grateloupia is often mistaken for dulse because both have similar shape and color, although *Grateloupia* has a slippery rather than leathery texture. Its blades range from pink to maroon in color, and it can be found in three forms: small branching blades at the base of a long narrow blade with ruffled edges; long blades connected end-to-end; or short and broad with a heart-shaped base.

Relationship to People

Grateloupia is an invasive species believed to have been introduced into Narragansett Bay around 1996 from Japan via spores in ship ballast water. Narragansett Bay was the first place that this algae was found on the East Coast of the United States.

This seaweed is commercially harvested in some regions for food consumption — the carrageenan extract helps thicken ice cream and toothpaste and is also used to make agar.

FIELD MARKINGS:
Blades range from pink to maroon and are thin and slippery. Found in three forms: small branching blades at the base of a long narrow blade with ruffled edges; long blades connected end-to-end; or short and broad with a heart-shaped base. Size: up to 3 feet long, although it can grow up to 9 feet.

HABITAT:
Attached to hard substrate in the lower intertidal to the subtidal zone.

SEASONAL APPEARANCE:
Year-round (in crust form for overwintering).

SENSITIVITY LEVEL:

Irish Moss

SCIENTIFIC NAME:
Chondrus crispus

Distinguishing Features and Behaviors

Irish moss is a red seaweed with flattened blades that fork off from a short stalk to form fingers with round, blunt tips. The blades taper towards disk-shaped holdfasts, which are actually rootlike anchoring structures. Typically, the blades are flat and slender, but in shallow water, they spread and take on a broader form. In deeper, subtidal zones, Irish moss blades are more narrow.

Irish moss is usually reddish-brown, but may also be white, brown or green. If washed ashore, Irish moss will bleach to white. It grows between two to six inches high and four inches wide and is usually found in clumps.

Irish moss can withstand extremely cold water temperatures and can survive being frozen during the winter months.

Relationship to People

Irish moss is found in the middle of the intertidal zone, below the rockweeds and above the kelp. It's less slippery than other intertidal seaweeds, allowing for safer footing on rocks than rockweeds and kelp provide.

Irish moss is used as a thickener in soups and dairy products and is harvested in New England to produce an extract called carrageenan. Carrageenan is a gelatinous carbohydrate used to emulsify dairy products, baked goods and cosmetics. It is also used to clarify beer, tan leather and package fish and meat.

This plant earned the name "Irish moss" because it grows in Ireland, where it was used in cooking and as a medicine long before Europeans introduced it to Narragansett Bay.

FIELD MARKINGS:
Reddish-brown seaweed with flattened blades that fork off into rounded tips. Size: 2 to 6 inches tall and 4 inches wide.

HABITAT:
Inlets, tide pools, lower intertidal zone.

SEASONAL APPEARANCE:
Year-round.

SENSITIVITY LEVEL:

Common Reed

COLLOQUIAL NICKNAMES:
Phragmites, Giant Reed

SCIENTIFIC NAME:
Phragmites australis

Distinguishing Features and Behaviors

Phragmites is a tall, stately member of the grass family. This species stands five to fifteen feet high and produces large purple or tan plumelike flowers six to twelve inches long that bloom in August. As they age, the flowers turn light brown and feathery. The mature flowering bodies are present on the reed until the next flower blooms the following year. The leaves of *Phragmites* are smooth, flat and green; they can grow as large as 20 inches long and two inches wide.

Phragmites grows by sending out rhizomes — long root runners that spread underground outwards of 17 to 34 feet from the plant. These runners sprout frequently, producing large colonies of reeds and are often found adjacent to cattails.

The *Phragmites* most often seen in Rhode Island is not native and can be an aggressive colonizer of salt marshes. In many areas, it has replaced other tidal marsh grasses. Recently, a native species of *Phragmites* has been identified. This reed is generally found in less dense patches than the non-native reed and is identified by the bare stems of the previous year's growth. The non-native reed retains its leaf sheaths.

Relationship to People

Phragmites, or *Phrag*, as it's commonly called, is found everywhere in the world except for parts of Polynesia. It occurs naturally in the upland zone of many Narragansett Bay marshes, forming a line between fresh and brackish water. Following alteration by development and construction, *Phragmites* can take over a marsh, with its root system actually drying out the marsh.

Because cordgrass needs a wetter, undisturbed habitat to survive, *Phragmites* stands have replaced cordgrass in extensive areas of tidal marsh and coastal wetlands. Altered marshes where *Phragmites* has taken over, do not support as many different species as undamaged cordgrass marshes. Dense thickets of dried-out reeds can also create a fire hazard. Marsh restoration efforts often target areas where *Phragmites* is abundant.

FIELD MARKINGS:
The reed is light green in the spring and summer; light brown in the fall. Flowers are purple or tan, changing to light brown with age. Flower stalks remain over winter. Size: grows in dense stands 5 to 15 feet tall.

HABITAT:
Fresh to brackish marshes; more common along the coastline.

SEASONAL APPEARANCE:
Blooms in August.

SENSITIVITY LEVEL:

High Tide Bush

COLLOQUIAL NICKNAMES:
Marsh Elder

SCIENTIFIC NAME:
Iva frustescens

Distinguishing Features and Behaviors

High tide bush, also known as marsh elder, is a perennial, deciduous shrub found at the inland edge of salt marshes around Narragansett Bay. It gets its name from growing at the upper edges of the high tide line.

The succulent leaves are egg-shaped to narrowly lance-shaped and are oppositely arranged except for the upper reduced leaves. The lower leaves grow four to six inches long and one to two inches wide. Many greenish-white flower heads are arranged on terminal stems subtended by tiny green, leaflike appendages.

High tide bush naturally grows in the mid- to high salt marsh estuarine area or back dunes and on muddy sea shores from Massachusetts to Florida to Texas. It usually only occurs at elevations where its roots are not subject to prolonged water table flooding, such as the upland border of salt marshes.

Relationship to People

The high tide bush is normally associated with the mid-to high marsh ecosystem, forming the last line of defense for shoreline erosion control. More recently, this species has become associated with the lower marsh system by colonizing the dredged material resulting from construction of ditches and ponds for mosquito control.

FIELD MARKINGS:
The shrub bears purple to brownish fruit with a gray-brown bark. Evergreen, spear-shaped, three-veined leaves with sawlike edges are arranged opposite each other on the twigs.
Size: 4 to 6 feet.

HABITAT:
Upper salt marsh.

SEASONAL APPEARANCE:
Year-round; flowers in summer.

SENSITIVITY LEVEL:

Pickleweed

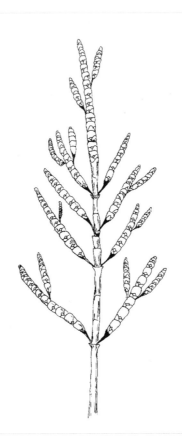

COLLOQUIAL NICKNAMES:
Glasswort, Sea Pickles,
Samphire

SCIENTIFIC NAME:
Salicornia spp.

Distinguishing Features and Behaviors

Pickleweed is a smooth, salt-tolerant plant common to the salt marshes of Narragansett Bay. The plant grows four to twenty inches tall and has jointed, branching stems that range in color from bright green to deep red. The leaves are scalelike formations along the segments of the stem.

Pickleweed is related to cacti and has fleshy, plump stems, resembling asparagus spears. These stems allow the plant to increase its water-holding capacity to survive in harsh, salty and dry conditions. The flowers of the pickleweed are small and green. In the fall, this plant turns bright crimson, adding dramatic beauty to the salt marshes it inhabits.

The segmented stems of the pickleweed are succulent because they contain so much water. Due to its fibrous and resistant root system, pickleweed is well adapted to a harsh existence in a salt marsh environment.

The plant is extremely salt-tolerant and is usually the first to colonize bare areas in high-saline marshes. These areas, called pannes, are so harsh they restrict the ability of most plants to survive in them. Pickleweed does not require much water because it stores needed supplies in its stems. It provides a habitat for some invertebrates and is a food source for many animals.

Relationship to People

The young stems of pickleweed are edible to humans. They have a salty taste and can be eaten pickled or as a garnish in fresh salads.

FIELD MARKINGS:
Low-growing, fleshy plant with jointed, branching stems that range in color from bright green to deep red. Size: 4 to 20 inches tall.

HABITAT:
Salt marshes, beaches.

SEASONAL APPEARANCE:
Year-round; blooms late August through November.

SENSITIVITY LEVEL:

Saltmarsh Cordgrass

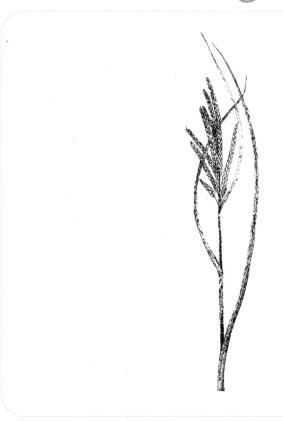

COLLOQUIAL NICKNAMES:
Saltwater Cordgrass, Marsh
Grass, Smooth Cordgrass

SCIENTIFIC NAME:
Spartina alterniflora

Distinguishing Features and Behaviors

Saltmarsh cordgrass is a tall, smooth grass ranging in height from six inches to seven feet. It grows in three different sizes depending on its location in the salt marsh: tallest near the water, of intermediate size behind tall saltmarsh cordgrass and shortest near the high marsh meadow grass. The shorter saltmarsh cordgrass can often be found in areas of low elevation in the marsh.

The flowering stalks resemble wheat and are arranged along one side of the stalk, similar to saltmeadow cordgrass. Its leaves are thick and wide, and the root structure is strong and complex.

Saltmarsh cordgrass is one of the most common forms of vegetation found in Narragansett Bay salt marshes and is a vital plant species in the estuary. Often only a small amount of saltmarsh cordgrass, called fringe, is found extending from the high marsh to the water. However, saltmarsh cordgrass also occurs in large fields, usually near the head of tidal creeks, and may be submerged at high tide.

This plant is important to marsh health due to the high volume of organic material it contributes during decomposition. In fact, saltmarsh cordgrass is the most productive of the marsh grasses. Located in low marsh areas, it is flooded twice daily by the tidal action of the Bay. The complex root system of the cordgrass helps bind it to the banks, preventing the tide from eroding the shoreline. Although it relies primarily on groundwater absorbed through the roots, cordgrass is able to extract fresh water from salt water when the need arises.

Relationship to People

Salt marshes serve as pollution filters and act as buffers against flooding and shoreline erosion. Many of Narragansett Bay's salt marshes have been severely degraded by filling, ditching, construction and runoff pollution. Construction of roads and filling of marshes can alter the flow of tidal waters, negatively influencing marsh ecology. This encourages the growth of invasive plants such as *Phragmites*, which outcompete the native marsh grasses for space. Salt marshes are increasingly stressed by rising rates of sea level due to climate change. Marshes that can't build up their elevation in time with sea level rise will likely drown.

FIELD MARKINGS:
Tall, smooth grass, green in spring and summer; turns light brown in late fall and winter. Flowering stalk visible in summer and fall. Size: 3 to 7 feet. A shorter form, 6 to 12 inches high, occurs in low-oxygen areas of the high marsh.

HABITAT:
Intertidal zone of salt marshes, banks of bays and creeks.

SEASONAL APPEARANCE:
Blooms July to August.

SENSITIVITY LEVEL:

Saltmeadow Cordgrass

COLLOQUIAL NICKNAMES:
Saltmeadow Hay,
Marsh Grass

SCIENTIFIC NAME:
Spartina patens

Distinguishing Features and Behaviors

Saltmeadow cordgrass is a slender and wiry plant that grows in thick mats one to two feet high. Its stems are wispy and hollow, and the leaves roll inward and appear round. Because its stems are weak, wind and water action can bend the grass, creating the appearance of a field of tufts and cowlicks. Mats of saltmeadow cordgrass are inhabited by many small animals and are an important food source for ducks and sparrows.

Like its relative saltmarsh cordgrass, saltmeadow cordgrass produces wheatlike fruits and flowers on only one side of the stalk. Flowers are a deep purple from June to October and turn brown in the winter months. Saltmeadow cordgrass is found in high marsh zones where it is covered at times by high tide. Specialized cells are able to exclude salt from entering the roots, preventing the loss of fresh water. This cordgrass is less tolerant to salt water than other salt marsh grasses.

A healthy salt marsh depends on the presence of both saltmeadow and saltmarsh cordgrass. Cordgrasses provide rich habitat for juvenile and adult crustaceans, mollusks and birds, and they serve as a major source of organic nutrients for the entire estuary.

Relationship to People

Salt marshes serve as pollution filters and buffers against flooding and shoreline erosion. During the colonial era, towns around Narragansett Bay were often settled based on their proximity to salt marshes. Saltmeadow cordgrass was harvested for bedding and fodder for farm animals, and for garden mulch. Before hay was baled and stored under cover, cordgrass was used to top the hay stacks in the fields.

Many of the salt marshes in the Bay have been severely affected by filling, development and road construction. These alterations restrict tidal flow, often having a severe impact on the marsh. Because saltmeadow cordgrass requires a salty, wet habitat, restricted tidal flow often dries out the marsh and encourages the growth of invasive plants. Cordgrasses are often outcompeted for space by common reeds in areas where human activity has disturbed or altered the marsh. Common reeds are not as productive or beneficial to a salt marsh as cordgrass.

FIELD MARKINGS:
Haylike grass found in the upper areas of the marsh. Green in spring and summer; turns light brown in late fall and winter. Size: 1 to 2 feet tall.

HABITAT:
High salt marsh zone.

SEASONAL APPEARANCE:
Blooms June to October.

SENSITIVITY LEVEL:

Sea Lavender

COLLOQUIAL NICKNAMES:
Beach Heather,
Marsh Rosemary

SCIENTIFIC NAME:
Limonium carolinianum

Distinguishing Features and Behaviors

Sea lavender is a flowering plant found in low tidal areas of salt marshes, usually interspersed with saltmarsh and saltmeadow cordgrasses. Its tall, thin stems often have a reddish tinge and grow six to twenty-four inches high. Due to its location, sea lavender is commonly submerged during high tide.

Sea lavender has dark green, leathery, spoon-shaped leaves that grow in a circular pattern at the base of its stem. Its branches produce small, fragrant, pale purple flowers along one side. Sea lavender flowers originate from basal leaves that rise up directly from the plant's roots. The leaves of the sea lavender grow upward along its stem in sheaths, forming a tubular envelope that surrounds the stem. The sheath causes the flowers to appear alternately and delicately arranged.

Sea lavender is a perennial and persists for years once it becomes established. The first sea lavender flowers can be seen in July, and flowering continues into October.

Relationship to People

Sea lavender is one of Narragansett Bay's most beautiful coastal plants. Unfortunately, coastal development has destroyed much of its habitat in past years. Also, because of its popularity in floral decorations, sea lavender has been overcollected. Picking it while in its flowering stage stops the plant from producing seeds for future generations; pulling it up by the roots destroys the entire plant.

In May 1991, sea lavender was added to a Rhode Island state law that protects many wild, decorative plants. As a result, it is illegal to dig or pick sea lavender or other plants protected by this law, unless you are on your own property or have the landowner's written permission. All plants on town, state and federal property are protected, as are all plants on private conservation lands. Sea lavender is grown commercially by florists for flower decoration.

FIELD MARKINGS:
Dark green leaves surround the base of this perennial plant. Branching flower stalks support small, eighth-inch, delicate pale purple flowers. The entire plant turns brown in fall and winter. Size: 6 to 24 inches high.

HABITAT:
Low tidal salt marsh areas, irregularly flooded salt marshes, occasionally on coastal sand dunes.

SEASONAL APPEARANCE:
Blooms July to October.

SENSITIVITY LEVEL:

Spike Grass

COLLOQUIAL NICKNAMES:
Salt Grass

SCIENTIFIC NAME:
Distichlis spicata

Distinguishing Features and Behaviors

Spike grass is a short grass often found growing with saltmeadow cordgrass in higher salt marsh zones. The two are difficult to distinguish. When in bloom, spike grass has a small flower at its tip, known as the terminal spike.

Spike grass also has stiff, wiry, light green stems eight to fifteen inches high. Stems of this grass are usually hollow and erect, but winds and tides can bend them to form mats. Its leaves are also hollow, curl inward, appear round and radiate in several planes around the stem. The leaves can reach lengths of two to four inches.

Dead vegetation washed up by the tide creates bare patches in the marsh, a healthy and natural process. Dense clumps of dead vegetation prevent sunlight from reaching the sediment below, causing all of the vegetation under the clump to die. The depression left after the vegetation decays fills with water after each flooding tide. Periods of intense sunlight that dry out the sediment along with tidal water flow into the depression results in an area that is too salty and harsh for most plants to survive. These areas, called pannes, usually appear as bare dirt patches encrusted with white salt crystals.

Spike grass is a quick colonizer of bare patches in the marsh. Spike grass living in a healthy section of the marsh will send out root runners to the bare patches to expand and exploit the open space. The new shoot does not actually grow in the panne; it gets water and nutrients from the original plant. Over time the developing spike grass will create enough shade to reduce the impact of the sun on the sediment, allowing other marsh plants to recolonize the bare patch.

Spike grass provides a habitat for many invertebrates and is the base of the food chain, with the rootstalks and young plants providing food for waterfowl, marsh birds, sea birds and small mammals.

Relationship to People

A well-developed root system, high salinity tolerance and the ability to colonize bare marsh peat make spike grass an important marsh stabilizer. Spike grass also helps provide a shoreline buffer and filter polluted runoff. Damage to spike grass through human activity leads to shoreline erosion, reduces the natural pollution filter system of the salt marsh and destroys habitat.

FIELD MARKINGS:
Terminal flowering spike at its tip. Light green in spring and summer; becomes light brown in fall and winter. Size: 8 to 15 inches tall.

HABITAT:
High salt marsh.

SEASONAL APPEARANCE:
Blooms August to October.

SENSITIVITY LEVEL:

American Beach Grass

COLLOQUIAL NICKNAMES:
Beach Grass

SCIENTIFIC NAME:
Ammophila breviligulata

Distinguishing Features and Behaviors

American beach grass is a dominant plant on sand dunes bordering Narragansett Bay. Growing in tufts of up to 12 stems, the thick, green blades of beach grass can grow from two to four feet tall. Beach grass reproduces through seeds produced by the plant, but either a lack of fresh water or excessive sand burial just after germination can restrict plant growth.

Below the sand, beach grass is anchored by rhizomes, an underground network of stems and roots. Rhizomes can reach depths of six feet and help stabilize the dunes. With the assistance of beach grass, dunes grow and spread from their early formation stage and eventually fuse together to create sand dunes. The size and shape of the dunes are directly affected by the plant's ability to "capture" the sand. Beach grass leaves act above-ground while rhizomes act below-ground to keep the sand from blowing or eroding away. Beach grass contributes to the high growth rate of sand dunes, often adding nearly ten inches of sand per year.

Another important dune species is the beach pea, *Lathyrus japonicus*. The compound leaves of the beach pea have six to ten oval leaflets, and the whole leaf bends in an arc. Its flowers are pink or purple. The beach pea also helps stabilize dunes with its extensive root and rhizome system. As a nitrogen fixer, the beach pea provides nutrients to other dune species, including beach grass.

Dunes provide habitat for many animals, including nesting areas for the piping plover, an endangered species of sandpiper that colonizes certain barrier beaches.

Relationship to People

Beach grass can be damaged simply from people and domesticated animals walking or running over it. The destruction of beach grass leads to beach and dune erosion and loss of habitat for nesting birds. Boardwalks and trails have been constructed to protect beach grass, and, ultimately, help prevent dune erosion.

FIELD MARKINGS:
A perennial flowering grass with stiff, sharp, dark green blades that turn brown in fall and winter. Size: 2 to 4 feet tall.

HABITAT:
Sandy beaches, coastal dunes.

SEASONAL APPEARANCE:
Blooms July to September.

SENSITIVITY LEVEL:

Bayberry

Detail of leaf

COLLOQUIAL NICKNAMES:
Northern Bayberry

SCIENTIFIC NAME:
Myrica pensylvanica

Distinguishing Features and Behaviors

Bayberry is a many-branched shrub found commonly in uplands near the coast. It has deep green alternate leaves that are typically clustered toward the ends of the twigs and are usually between one and one half to four inches long. The leaf shape is elliptical and often appears slightly twisted with a wedge-shaped base and, occasionally, a few small teeth near the tip of the leaf. The leaves can either be dull or glossy, with resinous glands that produce an oil, which gives the bayberry its distinctive fragrance.

The bayberry plant is dioecious, meaning that male and female flowers grow on separate plants. The round fruits that form on the female plant are initially green but turn dusky blue with a bumpy surface.

Bayberry plants grow well in harsh coastal conditions, including sites with high winds, dry soils and salt spray.

Beach plum *(Prunus maritima)* is another coastal upland shrub occasionally seen in Rhode Island, though more common along the southern shore.

Relationship to People

The wax on bayberry fruits is used traditionally to make bayberry candles. Forty pounds of berries produces one pound of wax. The berries are boiled in water until the wax floats to the surface. The hot water and hot wax are strained through a cloth to remove any solids, then wicks are dipped into the water and wax to create the candles. Because of the intensive effort involved in creating bayberry candles, they are quite expensive and considered a luxury.

FIELD MARKINGS:
Deep green leaf, elliptical in shape. Female green fruits develop in March but become gray and waxy. Size: averages 3 to 4 feet tall, but can grow up to 10 feet.

HABITAT:
Wet or dry acidic soils, sand dunes, coastal shrubs, open woods, upper edge of salt marshes and moist shores.

SEASONAL APPEARANCE:
Blooms in early spring.

SENSITIVITY LEVEL:

Beach Rose

Colloquial Nicknames:
Sea Rose, Salt-spray Rose

Scientific Name:
Rosa rugosa

Distinguishing Features and Behaviors

Beach rose is a flowering plant common along the coastline of Rhode Island. It grows in large shrub groupings that stand three to six feet tall, often form-ing thick hedges. The branching stems are dense and bristly, covered in sharp thorns and produce dark green, shiny compound leaves three to six inches long.

Commonly found along many Narragansett Bay beaches, beach roses range across North America from Nova Scotia to Florida and along the Gulf Coast.

Blooming in the summer, the beach rose produces two- to four-inch flowers ranging in color from white to pink to deep rose. After blooming, the flowers bear fruit called rose hips, which last through summer and into the winter months. The fruit is orange to brick red and is eaten by birds and other animals, including people. Rose hips resemble small crab apples and have a slightly bitter taste. The rose hip contains an abundance of seeds, which are distributed, in part, through the waste of the animals that feed on them. The rose hip also floats — this waterborne method of distribution is part of the reason that beach roses are considered extremely invasive.

Like American beach grass, the beach rose thrives in sandy habitats. Its dense thickets are inhabited by small animals and used as nesting sites by many birds.

Relationship to People

The beach rose is not native to the United States. It was brought over from eastern Asia in the mid-1800s for use as an ornamental plant. The fruit of the beach rose is high in vitamin C and can be eaten raw, used in jelly or made into tea. Rose hip tea is believed to aid in circulation and help digestive disorders.

Beach rose is an invasive shrub, and its planting is not recommended. Try planting Virginia rose *(Rosa virginiana)* instead; it has characteristics similar to beach rose and is native.

FIELD MARKINGS:
Dark green bush.
Blooms range in
color from white to
pink to deep rose.
Size: 3 to 6 feet tall.

HABITAT:
Coastal sand dunes,
salt marshes, rocky
shores, roadsides.

SEASONAL APPEARANCE:
Blooms May to
September.

SENSITIVITY LEVEL:

Ocean

 Estuary

 Shoreline

WHERE DO I FIND IT?
Use these icons as a quick reference to where you might find a particular Bay species. Icons denote a species' predominant habitat; other preferred habitats are mentioned within the text.

Invertebrates

Narragansett Bay may be famous for its quahogs and shellfish, but there are many other invertebrates that play a critical role in the balance of life in the Bay.

Invertebrates are animals without a backbone, and they make up the majority of species in the animal kingdom. In the *Uncommon Guide*, inverts are divided into sponges, jellies, segmented worms, arthropods, mollusks and echinoderms. Inverts live in all-water body zones; some are benthic (live on the bottom), some are nekton (free swimming in the water column) and some are planktonic (live in the water column but mostly float and drift).

Sponges have the simplest body plan of the multicellular inverts found here. With no true organs or tissues, they act as a colony of individual cells. Jellies comprise two groups. True jellies use stinging cells (cnidocytes), whereas comb jellies use sticky cells (colloblasts) to capture prey. True jellies are in the phylum that includes jellyfish, coral and anemones. Comb jellies found in Narragansett Bay include the sea gooseberry and sea walnut.

Segmented worms, or annelids, are characterized by body segments that repeat most internal and external parts. Some annelids reproduce by a curious method called epitoky, where some or all of the segments are turned into reproductive cells and completely split apart.

Arthropods have jointed legs, segmented bodies and hard exoskeletons. The exoskeleton prevents water loss and adds a measure of protection, but it does not grow with the animal; it must be shed through molting. The majority of marine arthropods are crustaceans, which include shrimp, lobsters and crabs.

Mollusks are animals with a thick, muscular foot for movement, a covering (or mantle) that protects the animal and a drilllike tongue (called the radula) used for feeding. One-shelled mollusks are called gastropods; two-shelled mollusks are known as bivalves. Squid and octopus belong to a division called the cephalopods. Mollusks include oysters, clams, mussels, scallops and snails.

Echinoderms are symmetrical animals with thick, spiny skins. They have a unique water vascular system that controls their movement, feeding, gas exchange and sensory perception. Included in this phylum are sea urchins, sea stars and sea cucumbers. Other invertebrate groups include bryozoans, which are sessile and colonial, and tunicates, which are actually somewhere between inverts and chordates due to the presence of a primitive notochord (a spinal cord precursor) during development.

Boring Sponge

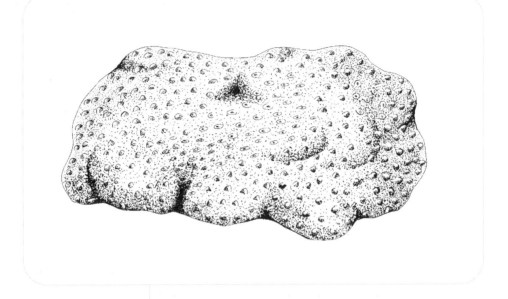

COLLOQUIAL NICKNAMES:
Sulfur Sponge,
Monkey Dung

SCIENTIFIC NAME:
Cliona celata

Distinguishing Features and Behaviors

When clustered in a colony, the boring sponge resembles a large, firm, irregularly shaped household sponge. Covered by small pores, boring sponges are simple animals belonging to the phylum Porifera, meaning "pore-bearers." The sulphurous-yellow pores of the boring sponge protrude from holes in mollusk shells or coral. In some cases, they may actually grow over the host entirely.

Considered among the most simple and primitive of all multicellular animal groups, sponges lack specialized organs and distinct tissues. Basic life functions such as eating, breathing and removal of waste are carried out exclusively by individual cells acting independently of one another. As filter feeders, boring sponges eat small particles of food brought into their bodies via sea water. They have no nervous system and are non-responsive to touch.

The larvae of boring sponges settle onto shells of oysters and other mollusks. After settling, they develop into tiny sponges and burrow into the host by secreting sulfuric acid. The acid chemically etches small tunnels and slowly eats away at the shell, weakening and even disintegrating the host.

Oyster and clam shells covered with small pocklike marks show evidence of boring sponge infestation. Shells occupied by boring sponges are brittle and breakable. Sometimes the boring sponge colonies will entirely cover a shell in a thick orange mass, often killing the oyster or clam. A healthy oyster can combat the attack by filling in the holes left by the sponge with a thin layer of shell material. Oyster and clam shells damaged by these holes are usually so weakened that they dissolve, which reduces shell accumulation on the Bay's bottom.

Relationship to People

Boring sponges are considered a nuisance by commercial fishermen. Their large numbers present in the catches of trawling vessels result in an enormous amount of excess weight.

FIELD MARKINGS:
Bright yellow to orange-red. Size: individuals are half an inch wide and one-sixteenth of an inch tall. Colonies can be up to 12 inches long.

HABITAT:
Along the rocky bottom and on shells of dead mollusks and corals.

SEASONAL APPEARANCE:
Year-round.

SENSITIVITY LEVEL:

Red-beard Sponge

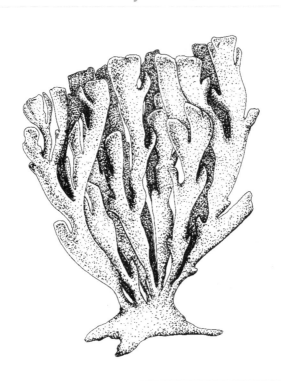

Colloquial Nicknames:
Red Sponge

Scientific Name:
Microciona prolifera

Distinguishing Features and Behaviors

Sponges are colonies made up of many individuals living within a hard, shelllike capsule. They are primitive animals, not plants, and can be found in a variety of forms and habitats. Sponges lack true tissues and organs.

Red-beard sponges — common to Narragansett Bay — begin as thin encrusting layers and eventually develop into heavy masses. Although species of sponges are difficult to tell apart, red-beard sponges are the only ones in the Bay with thick and fleshy, entwined fingerlike branches.

A firm exoskeleton that consists of pores used for feeding and respiration surrounds the fleshy material of the body. Water is brought into the hollow cavity of the sponge through cilia-lined pores. As water passes through the pores, the cilia trap oxygen and planktonic organisms for breathing and eating, respectively. Once the nutrients are removed, water and waste products exit the sponge through an opening at the top.

Sponges are filter feeders that cannot live out of water and are, therefore, most common in areas where constantly submerged. Eggs develop within the sponge itself and are released as free-swimming larvae. Once it has undergone a planktonic stage, the sponge will attach itself to a substrate and become stationary.

Relationship to People

Shrimp, worms, mud crabs and some species of small fish take refuge inside the folds of sponges. Although Narragansett Bay is home to 16 species of sponges, the red-beard is the predominant species. They are pollution-tolerant and can thrive in the low salinities of estuarine waters.

Red-beard sponges can be destructive to oyster beds. Using oyster shells as attachment sites, the red-beard sponge can suffocate or erode the shell of the oyster.

FIELD MARKINGS:
Distinctively bright red or orange. Size: up to 12 inches wide and 8 inches high.

HABITAT:
Attached to rocks, pilings or other hard objects in shallow bays, salt marshes or estuaries.

SEASONAL APPEARANCE:
Year-round.

SENSITIVITY LEVEL:

Comb Jellies

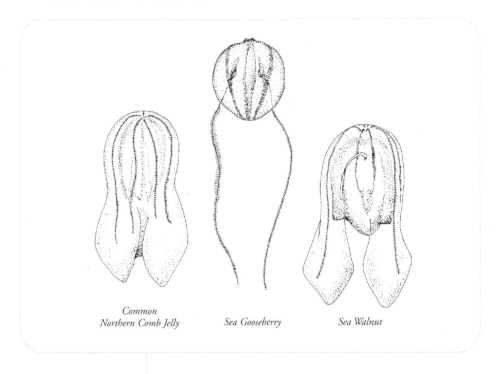

Common
Northern Comb Jelly   Sea Gooseberry   Sea Walnut

COLLOQUIAL NICKNAMES:
Common Comb Jelly,
Sea Gooseberry, Sea Walnut,
Jellyfish

SCIENTIFIC NAME:
Bolinopsis infundibulum,
Pleurobrashia pileus,
Mnemiopsis leidyi

Distinguishing Features and Behaviors

Comb jellies, or ctenophores, are transparent, delicate, gelatinous creatures. Composed of protein, salt and 95% water, they are difficult to preserve as their bodies break apart. They are best observed underwater, in their natural environment.

The comb jelly is separated into sections by ridges or bands of cilia arranged in rows along the body. These bands, called comb rows, are used for forward and backward locomotion when the comb jelly beats them together. Considered part of the planktonic community, comb jellies are poor swimmers and are at the mercy of the tides and currents, constantly being washed, blown or driven ashore in stormy seas. Comb jellies do not have the stinging tentacles common to other jellies in the Bay.

There are several hundred different species of comb jellies in the oceans of the world, many of which are found in Narragansett Bay. Common northern comb jellies – the most common in the Bay – are somewhat flattened and oval, with lobes shorter than the length of their body. Sea gooseberries, generally found in large swarms, are round with a pair of extendible tentacles less than an inch long. Sea gooseberries turn a brilliant scarlet red when threatened. Sea walnuts measure about four inches long, are bilobed, luminescent at night and have lobes extending past the length of their bodies.

Due to their transparency, comb jellies are difficult to see in daylight. At night, luminescence from the action of the comb rows causes them to glow a soft green color when the water is disturbed. These creatures feed voraciously on plankton by opening their mouths while swimming.

Relationship to People

The feeding habits of comb jellies can have a serious effect on the Bay ecosystem. A swarm of comb jellies can devour large quantities of zooplankton living in an area. During times when the comb jelly population is high, they can be a problem for commercial oyster fisheries because the jellies prey heavily upon oyster larvae.

Comb jellies are tolerant of pollution and can be indicative of poor water quality.

Field Markings:
Translucent with pinkish-white feeding appendages. Size: 2 to 8 inches long.

Habitat:
The planktonic zone of shallow and open water.

Seasonal Appearance:
Year-round; most common in spring and summer.

Sensitivity Level:

Lion's Mane Jelly

COLLOQUIAL NICKNAMES:
Red Jelly

SCIENTIFIC NAME:
Cyanea capillata

Distinguishing Features and Behaviors

The lion's mane jelly can grow to be the largest jelly in the world. It is composed of a large, semi-transparent disk and an umbrella-shaped, lobed underside. Sixteen notches at regular intervals are present around the edge of the disk; eight of these are sensory organs. The lion's mane has no distinct head or centralized nervous system. This jelly is aptly named — its bushy, yellow-orange tentacles resemble the mane of a lion.

On the lower surface of the disk, a central mouth is surrounded by veillike lips and eight clusters of tentacles covered by stinging cells called nematocysts. The lion's mane jelly uses the nematocysts to sting and paralyze its prey, as well as to defend itself against predators. Long feeding arms extend below the bell of the lion's mane, trapping small fish and invertebrates as it drifts along. These arms also carry food to its mouth.

The umbrella-shaped body and bell of the lion's mane houses a powerful muscle that forces out water, helping the creature to move. 95% of its bulk is seawater. The rest of the mass is a combination of minerals and organic material, which forms the strong and resilient jellylike substance. Young fish, such as haddock and butterfish, are immune to the stinging cells and often live underneath the bell, traveling with the jelly for protection.

Relationship to People

Lion's mane jellies are common to Narragansett Bay in the late spring and early summer. The lion's mane is poisonous due to the paralyzing toxin in the nematocysts. The toxin is not present in strong doses, as it is only needed to stun small fish. Still, the lion's mane should be approached with caution, as the sting can be dangerous. If stung, people often feel numbness, a burning sensation or an inflammation of the skin, depending on the amount of toxin. Although a rare occurrence, this species can be fatal to swimmers who are allergic to the toxins. Lion's mane jellies often wash up along the beaches and should be cautiously avoided if walking barefoot along the shore.

FIELD MARKINGS:
Pink to brownish-purple bell, with yellowish tentacles that darken with age. Size: 6 inches to 3 feet in diameter, with tentacles from 1 to 5 feet long. Can reach sizes of 8 feet in diameter, with tentacles 200 feet long.

HABITAT:
Open water, bays, harbors, inlets.

SEASONAL APPEARANCE:
Spring and early summer.

SENSITIVITY LEVEL:

Moon Jelly

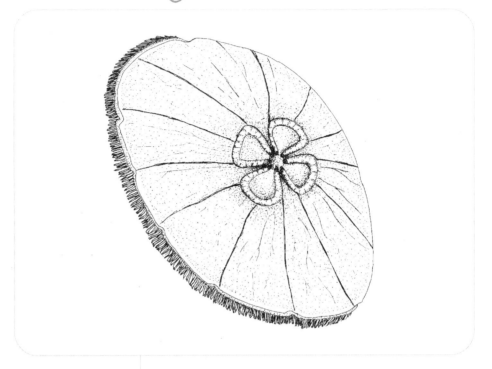

COLLOQUIAL NICKNAMES:
Crystal Jellyfish,
Saucer Jelly

SCIENTIFIC NAME:
Aurelia aurita

Distinguishing Features and Behaviors

The moon jelly is one of the most common jellies in Narragansett Bay. Its body shape resembles an umbrella, with four equal horseshoe-shaped stomachs in the center of its body. Moon jellies are made up almost entirely of water and will not hold their shape out of water.

Instead of long tentacles characteristic of other jelly species, the moon jelly has many relatively short tentacles. Four elongated feeding appendages, called "oral arms," surround the mouth.

To trap plankton, the moon jelly uses the underside of the umbrella, or bell, which is covered with a mucus lining. The oral arms are used to scrape its prey off the mucus lining and carry it into the mouth. This method of feeding is known as "suspension feeding."

"Jellyfish" is a misnomer commonly used for all jellies even though they are not related to fish. Jellies are considered to be among the largest members of the Bay's planktonic community. They are categorized as plankton because they don't have any method of locomotion beyond raising and lowering themselves in the water column. As such, they are at the mercy of tides and currents.

Relationship to People

The clear blobs often found washed ashore are most often dead moon jellies. The toxins in their tentacles are mild but can irritate the skin, causing a minor rash. In some countries, moon jellies are considered a delicacy and are widely consumed.

This jelly is a food source for many marine animals, including all species of endangered sea turtles. Plastic bags washed into Narragansett Bay are often mistaken for jellyfish by other marine life and may be ingested, resulting in suffocation or death.

FIELD MARKINGS:
Translucent white with darker tentacles and four yellowish-pink stomachs. Size: up to 12 inches in diameter.

HABITAT:
Open water, near the surface, close to shore.

SEASONAL APPEARANCE:
Late summer to mid-fall.

SENSITIVITY LEVEL:

Clamworm

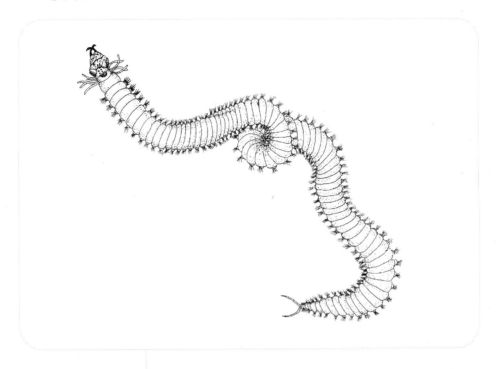

COLLOQUIAL NICKNAMES:
Sandworm, Ragworm

SCIENTIFIC NAME:
Nereis succinea

Distinguishing Features and Behaviors

Segmented, or annelid, worms are among the most unique and diverse worm species. This group includes the terrestrial earthworm and aquatic leech, as well as many marine worms — both sessile and free swimming — including bloodworms.

The clamworm belongs to the free-swimming group of segmented worms. The body of the clamworm consists of many segmented sections, each with a pair of bristly, red, paddlelike appendages. These paddles have gills used for respiration; they also allow the clamworm to swim or crawl along the sea floor.

Clamworms form transparent, mucus-lined sand burrows where they hide during daylight hours and come out at night to feed. A wandering, nocturnal hunter and a swift, voracious predator, the clamworm's keen sense of smell enables it to find prey.

The head of the clamworm houses sense organs, including four black eyes, which register changes in water chemistry and react to light and touch. The clamworm's proboscis is equipped with a set of jaws that open and shut through quick body movements. As soon as food is detected, the clamworm will thrust its mouth out and grab the prey with its powerful jaws. The clamworm feeds on other worms as large as or larger than itself. It also eats algae, small crustaceans, invertebrates and mollusks, including clams. Clamworms, in turn, are a significant food source for bottom-feeding Bay creatures and are preyed upon by horseshoe crabs, winter flounder, tautog and scup.

During mating season, the clamworm releases a segment of its body that contains gametes, or sex cells. These cast-off appendages can often be seen swarming in groups at the surface of near-shore waters during the spring and summer months. The adult worms remain on the bottom and die after spawning. These spawning aggregations attract schools of fish — striped bass in particular — which prey upon the worms.

Relationship to People

The clamworm is the most common and largest of the marine annelids living in the subtidal zone of Narragansett Bay. Some fishermen say that clamworms are the best bait for certain types of fish like winter flounder. These worms should be handled carefully since they can deliver a painful bite.

FIELD MARKINGS:
Iridescent green-blue or gray-brown, usually with fine red, gold or white spots. Size: 2 to 36 inches long.

HABITAT:
Rocky intertidal zone, mud and sand flats; often found under rocks.

SEASONAL APPEARANCE:
Year-round.

SENSITIVITY LEVEL:

Common Bamboo Worm

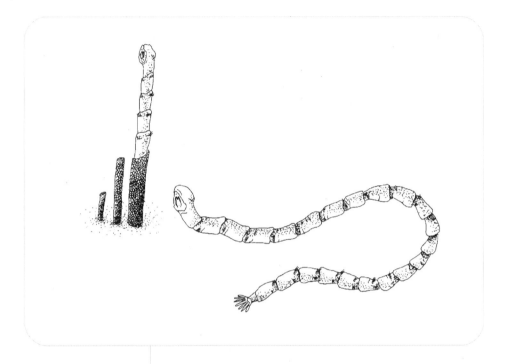

COLLOQUIAL NICKNAMES:
Bamboo worm

SCIENTIFIC NAME:
Clymenella torquata

Distinguishing Features and Behaviors

While several species of bamboo worm are found in the waters of the North Atlantic, the common bamboo worm is the largest and most abundant tube-building worm in Narragansett Bay.

These worms have thin, segmented bodies that resemble a stick of bamboo. Bamboo worms have no appendages or tentacles on the head; a hard plate at the tip of the head seals the tube when the worm retracts for protection. The tail is funnel-shaped, and the segments of the body are lined with hairlike bristles that anchor the worm into the tube. Bamboo worms live upside down in tubes constructed of grains of sand and cemented together with mucus produced by the worm.

Bamboo worms are detritus feeders. They feed by ingesting sand and mud, extracting detritus particles as they burrow. After digesting the food particles, excess mud and sand are excreted by the worm through the top of the tube.

Tube-building bamboo worms are visible in mud flats throughout Narragansett Bay. The tubes extend a few inches above the surface of the mud and generally occur in clusters of several individuals. The clusters provide sanctuary and habitat for other Bay life, including amphipods, isopods, mud snails and small crustaceans.

Relationship to People

Bamboo worms are also found in eelgrass beds. Human disturbance through nutrient enrichment and increased water temperatures from climate change reduces eelgrass habitat for these worms and other organisms that depend it for survival.

Environmental disasters, such as the 1996 North Cape oil spill off Moonstone Beach, can have a severe impact on the benthic community. Toxic oil will often kill all of the invertebrates and plant species it comes in contact with, sterilizing the entire area.

FIELD MARKINGS:
Light brownish-red to brick red. Size: up to 6.5 inches long.

HABITAT:
Sand or mud flats, eelgrass beds, intertidal zone to offshore.

SEASONAL APPEARANCE:
Year-round.

SENSITIVITY LEVEL:

Mud Worm

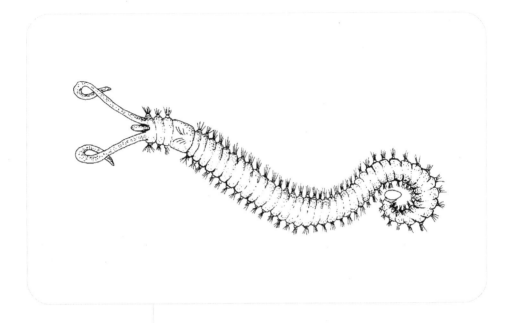

COLLOQUIAL NICKNAMES:
Whip Mudworm

SCIENTIFIC NAME:
Polydora ligni

Distinguishing Features and Behaviors

Mud worms are small, segmented tube-building worms similar to clamworms. They can be identified by two long, coiled organs, called palps, that extend from the head section and are used to sense and collect food. The fragile palps are often destroyed during collection of the worm, making identification difficult. Along each body segment, mud worms have bristles, or paddles, which are used for swimming and extra traction in the mud. Mud worms are slender, cylindrical and mostly transparent. They have two long antennae, a forked lobe above the mouth and a forked tail. Four eyes are positioned in a rectangular shape on its head.

Unlike wandering worms that seek out prey (such as the clamworm), mud worms capture food as it comes to them, using their specially modified palps. The fifth body segment is wide and has minute hooks and a suctionlike tail that holds the worm in its tube. The palps can often be seen extending from the tube of the worm as the animal grasps for its food.

Mud worms are among the most abundant worms in Narragansett Bay. Often gathering in large colonies, they can smother other benthic invertebrates in the area.

To collect food, the mud worm sweeps the bottom of the sea floor with its palps. They usually collect bits of sediment and excrete the mud after separating out the food. This mud will build up around the worms, creating a several-inch deep layer of sediment around the worm colony.

Relationship to People

In areas heavily populated with mud worms, over 50 worms can be present per square inch. This can have a significant impact on oyster beds, as the oysters can be completely buried in sediment accumulated from mud worms. This can lead to suffocation of the oyster bed.

Mud worms have been found in large numbers thriving in areas with considerable oil pollution. Experiments have shown that the populations will continue to grow even in the presence of oil.

FIELD MARKINGS:
Slightly reddish in color. Size: 1 inch long, quarter-inch wide.

HABITAT:
Intertidal and subtidal, rocky shores, soft mud or sand, on oyster beds, pilings.

SEASONAL APPEARANCE:
Year-round.

SENSITIVITY LEVEL:

Amphipods and Isopods

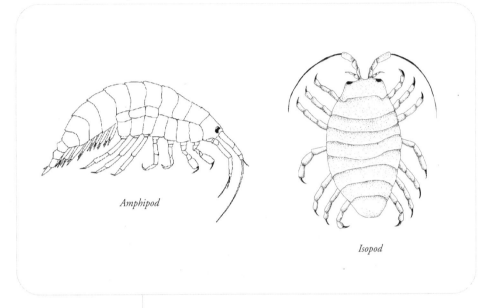

Amphipod

Isopod

COLLOQUIAL NICKNAMES:
Beach Fleas, Gribbles,
Sea Pill Bug

SCIENTIFIC NAME:
Amphipods: *Gammarus* spp.,
Talorchestia spp.
Isopods: *Sphaeroma* spp.,
Idotea spp., *Erichsonella* spp.,
Limnoria spp.

Distinguishing Features and Behaviors

The small flealike creatures seen scuttling around seaweed on the beach, burrowing in the sand, under rocks and in tidepools, generally belong to two groups of crustaceans called amphipods and isopods. Both amphipods and isopods belong to a group known as arthropods, which includes animals with jointed appendages and hard exoskeletons, such as crabs, lobsters, insects and spiders. The exoskeleton consists of armorlike, overlapping plates.

There are many different species of amphipods, and most are difficult to distinguish from one another. The name "amphipod" means "double," for two kinds of legs. Of their eight pairs of legs, the first five are used for walking and the last three, in the tail region, are modified for swimming. Some amphipods also have modified tail appendages for jumping. Most amphipods have several pairs of antennae, and some have an appendage used for grasping. Amphipods generally swim on their sides. Their bodies are flattened sideways with highly arched backs.

In contrast, isopods are flattened from top to bottom and are bottom dwellers. They are buglike in appearance, resembling insects found on land. Isopods are larger and are commonly found crawling and swimming among weeds, eelgrass, tide pools, dock pilings and rocks.

Both amphipods and isopods eat detritus (dead and decaying algae and seaweed) as well as other plants and animals. Decomposition of detritus is continued by bacteria and smaller animals that consume the waste of amphipods and isopods. Some of these creatures are parasitic on other Bay animals. Parasitic amphipods can be found on jellies, while parasitic isopods can live in the gills and fins of fish.

Relationship to People

Found throughout Narragansett Bay and along the eastern seaboard, amphipods and isopods are an important part of the food chain, playing a major role in the diet of various birds, fish and crabs.

During the summer, beachgoers commonly see large numbers of amphipods and isopods feeding on seaweed that has been washed ashore by wave action.

FIELD MARKINGS:
Amphipods range from gray to sand-colored to brownish-green to mottled with red or orange. Size: quarter-inch to 1 inch long.

Isopods are dull gray to yellow-brown. Size: quarter-inch to 2 inches long.

HABITAT:
Sandy beaches, seaweed, tide pools, tidal marshes above the high tide line, eelgrass beds, seagrass meadows.

SEASONAL APPEARANCE:
Year-round.

SENSITIVITY LEVEL:

Asian Shore Crab

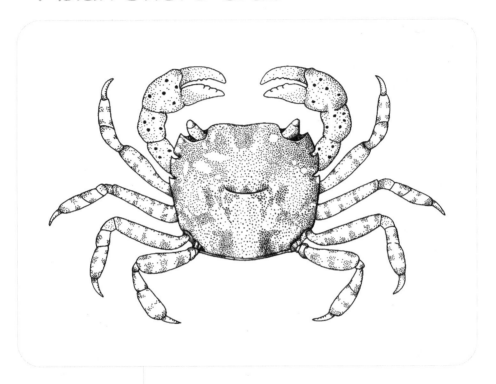

COLLOQUIAL NICKNAMES:
Japanese Crab

SCIENTIFIC NAME:
Hemigrapsus sanguineus

Distinguishing Features and Behaviors

The Asian shore crab has a square-shaped shell with three spines on each side. It is an omnivore, feeding on algae, marsh grasses and other invertebrates.

Females — who begin reproducing at a very young age — produce up to four clutches of 50,000 eggs, which are transported great distances in the water column. The Asian shore crab is eaten by gulls, seabirds, tautog, striped bass and other fish.

Relationship to People

As the name implies, this species is native to Asia and
arrived in New Jersey from Japan in the late 1980s.
It now ranges from the Carolinas to Maine. It was first
recorded in Narragansett Bay in 1993 by Save The Bay's
BayKeeper John Torgan and has since become
abundant on the rocky shore.

Asian shore crabs are now among the most abundant
crabs along the rocky shore of Narragansett Bay, from
the ocean up to Providence. They may be found easily
(sometimes by the hundreds!) by turning over rocks,
and they often exist alongside other native crab
species, such as the black-fingered mud crab.

Invasive species often cause environmental damage
by crowding out native species and out-competing
them for food and space. The exact environmental
impact of the Asian shore crab is unknown, but it is
being monitored closely in the Bay.

FIELD MARKINGS:
Mottled brown to
purple with tiger-
like bands on the
legs. Size: carapace is
about 1.5" across.

HABITAT:
Rocky intertidal from
the ocean to the head
of the estuary.

SEASONAL APPEARANCE:
Year-round.

SENSITIVITY LEVEL:

Atlantic Horseshoe Crab

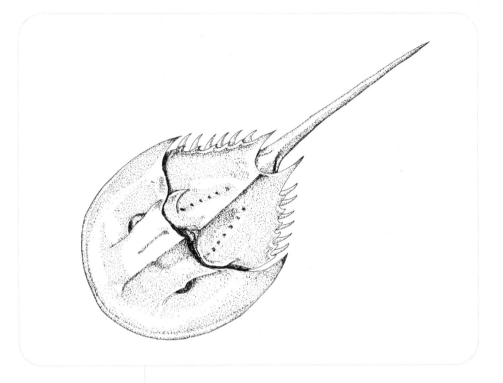

COLLOQUIAL NICKNAMES:
Horseshoe Crab, King Crab

SCIENTIFIC NAME:
Limulus polyphemus

Distinguishing Features and Behaviors

Horseshoe crabs are not true crabs. In fact, they are more closely related to spiders, ticks and scorpions. The brown horseshoe-shaped helmet of the shell protects the crab from sharks, turtles and seagulls. The spinelike tail of the horseshoe crab is not a weapon; it helps them turn over when they are upside down and acts as a rudder as they plow along the Bay bottom.

Horseshoe crabs walk with five pairs of legs, the last pair resembling a hand with five fingers. Unlike true crabs, the legs have long, slender claws that are not used for protection. The legs assist the horseshoe crab with eating by tearing and placing food into the centrally located mouth. These creatures feed on small clams and other bivalves, worms, invertebrates and organic compounds. Turning the crab over reveals the "book gills" — gills that resemble folded leaves of paper, which the horseshoe crab uses for breathing and swimming.

In late spring, adult horseshoe crabs migrate from deep water to mate along the shore where they gather in large groups. In place of the first pair of claws, the male has a modified hook used to grasp the female during mating. The crabs grow larger by molting — shedding the old shell and replacing it with a larger, soft shell from underneath that hardens in a few days. The shell left behind is often mistaken as a dead crab. Older crabs molt less frequently than juveniles and often become covered in algae and mollusks. The crabs reach maturity at nine to eleven years with some living more than 30 years.

Horseshoe crabs are considered to be "living fossils," and they are one of the most primitive crustaceans. They have existed in their present form since the Devonian period about 360 million years ago. Horseshoe crabs can live in extreme hot or cold temperatures and can even withstand being frozen in ice.

Relationship to People

Horseshoe crabs were once used primarily for fertilizer, chicken food and lobster bait. Today, they are still used as bait, but they also play a key role in medical research. Their blue, copper-rich blood contains the compound lysate, which is used in cancer research and the diagnosis of spinal meningitis.

The horseshoe crab's existence is threatened by pollution, destruction of habitat and increased harvesting for medical research.

FIELD MARKINGS:
Shell is light tan to dark brown. Legs and gills are dark reddish-brown with white markings. Size: averages 24 inches long and 12 inches wide. Females are larger than males.

HABITAT:
Salt marshes, ponds or estuarine waters with sandy or muddy bottoms.

SEASONAL APPEARANCE:
Year-round; more visible in spring when breeding.

SENSITIVITY LEVEL:

Black-fingered Mud Crab

COLLOQUIAL NICKNAMES:
Mud Crab

SCIENTIFIC NAME:
*Panopeus herbstii
and related species*

Distinguishing Features and Behaviors

The black-fingered mud crab is a small, mud-colored crab with large, black-tipped claws. The tips of their pinching claws are powerful enough to crack small, hard clam shells. The claws of the black-fingered mud crab are unequally sized, and the larger claw has a large tooth, which distinguishes this crab from others.

Mud crabs are generally found anywhere there is adequate protection. This includes pilings, sponges and bryozoans, oyster shells, rocks, soft mud and discarded cans and bottles. Mud crabs prey on young shellfish, mostly in oyster bed communities. Using their heavy, stout claw, they crush the shells of small mussels, barnacles and oysters. Mud crabs are eaten by birds and fish, such as cunner and tautog.

Several species of mud crabs are found in the Bay, differing mainly by the size and shape of the claws. They are more frequently found in submerged areas as the Asian shore crab has displaced the mud crab from its rocky shore habitat.

Another common species of mud crab is the white-fingered mud crab with its distinct white-tipped claws and white-striped legs.

Relationship to People

Mud crabs are not considered edible as they are too small to be worthwhile. Though powerful, their claws are too slow to be a danger to human fingers and toes.

Little is known about the mud crab population in Narragansett Bay, although they are commonly found living inside cans and bottles. The debris offers the little crabs protection and is a reflection of how these animals have learned to adapt to ever-increasing human impact. The white-fingered mud crab is not native to North America; they were accidentally imported in oysters from Europe.

FIELD MARKINGS:
Mud-colored with stout, black-tipped claws. Size: three-quarters of an inch wide and 1.25 inches long.

HABITAT:
Under stones and among the masses of sponges, seaweeds and discarded cans in muddy, brackish water; oyster beds.

SEASONAL APPEARANCE:
Year-round.

SENSITIVITY LEVEL:

Blue Crab

COLLOQUIAL NICKNAMES:
Blue Claw Crab

SCIENTIFIC NAME:
Callinectes sapidus

Distinguishing Features and Behaviors

Blue crabs belong to the family of swimming crabs that also includes the lady crab. The largest population of blue crabs are found in the Chesapeake Bay. Their name *Callinectes sapidus* comes from the Latin for "beautiful, savory swimmer." This crab has a characteristic sharp spine projecting outward from each side of its carapace. The rear pair of legs act as paddles, making these crabs excellent, rapid swimmers.

The blue crab has two stalked, compound eyes, which are controlled separately and can lay back into sockets in the shell. As predators, they will burrow in mud with only their eyes and antennae showing. Even the slightest shadow can trigger a reaction from the crab.

Blue crabs eat bivalves, crustaceans, fish, snails, live plants, dead animal material and even other blue crabs; their voracious feeding habits help regulate the populations of these species. They, in turn, are preyed upon by skates, striped bass, coastal birds, oyster toadfish, bluefish and sea stars during the crabs' dormant winter months. Blue crabs aggressively take on any opponent by raising their claws toward their enemies while scuttling sideways to escape.

Blue crabs grow by molting, or shedding their shells. They molt every few weeks when they are young but only once a year when they are older. Molting begins when a thin line appears down their backs. For adult crabs, molting occurs in the summer months, leaving them soft and defenseless for several days. At this point, they are known as "peelers" and are easy prey for wading birds, other crabs and mammals.

Relationship to People

Little is known about the blue crab population in the Bay, and although some years bring good numbers as far north as the Providence River, their total numbers are likely down from historic levels. Blue crabs are found from Narragansett Bay south to Florida, the Gulf of Mexico and the Caribbean.

A prized seafood in their soft-shell state, blue crabs can be cooked and eaten whole. They are harvested using dredges, crab pots, hand lines and dip nets.

Blue crabs need highly oxygenated water to survive. Nutrient pollution, including phosphorous and nitrogen from sewage plants, farm fields and lawns, causes algal blooms. When the free-floating algae die and decompose, large quantities of dissolved oxygen are consumed. Under these conditions, blue crabs will emerge from the water in large numbers to escape the deadly, deoxygenated water.

Field Markings:
Olive green carapace, with brilliant blue claws. Females have red-tipped claws.
Size: up to 9 inches long from point to point and 4 inches from head to tail.

Habitat:
Shallow and brackish waters, eelgrass beds, muddy bottoms.

Seasonal Appearance:
Early spring to late fall.

Sensitivity Level:

Fiddler Crab

Distinguishing Features and Behaviors

Fiddler crabs live in burrows near the water's edge. They are called fiddlers because of the extreme difference in the size of the male's claws, with the larger claw resembling a fiddle.

COLLOQUIAL NICKNAMES:
Sand Fiddler Crab,
Marsh Fiddler Crab,
Mud Fiddler Crab

SCIENTIFIC NAME:
Uca pugilator,
Uca pugnax,
Uca minax

All fiddler crabs are similar in shape, having a smooth carapace and a square-shaped body. The eyes are found at the end of two long and slender movable eyestalks located in the center of the carapace. The large second claw of the male fiddler crab is used to protect territories and to attract a mate during the breeding season. The male crab will stand by the entrance to the burrow waving the larger claw in an effort to attract a female.

Fiddler crabs are colonial, often living together in large clusters. Territorial fighting occurs between the males, and they will go to extremes to defend their burrows. Despite their fighting, fiddler crabs travel in herds of thousands when feeding. They live in slanting burrows up to three feet long, which they dig with their walking legs. The crab plugs the entrance when the tide rises, and remains in the burrow throughout the winter.

Fiddler crabs feed on plant detritus present in salt marshes after a receding tide. They emerge from their burrows in droves to filter through the mud. The fiddler crab can stay out of the water in damp ground for months at a time. They have gills for breathing in water, and they also have a primitive lung, which enables them to live on land.

Narragansett Bay also hosts a similar species called the marsh crab *(Uca pugnax)*. The marsh crab has two small, equal-sized claws and a square back with eyestalks on the outer corner of the carapace. Marsh crabs burrow with fiddler crabs and, although they are primarily herbivores, they sometimes prey on the fiddlers.

Relationship to People

Fiddler crabs play a vital role in salt marsh ecology because their feeding and burrowing keeps marshes clean and aids their growth. The long, hollow burrows of a fiddler crab community help to aerate the sediment of the marsh. Destruction of fiddler crab communities by shoreline development or other human activities also damages vital Bay salt marshes and wetlands.

FIELD MARKINGS:
Male fiddler crabs have one large claw and are brighter in color, having a purple-gray or blue carapace with irregular markings of black or brown. Females have equal-sized claws and much more subdued coloration on their carapaces. Size: carapace is up to 1 inch long; major claw on the male can be up to 2 inches long.

HABITAT:
Mud, sand or marsh, near the high tide line.

SEASONAL APPEARANCE:
Year-round; hibernate in burrows during the winter months.

SENSITIVITY LEVEL:

Green Crab

COLLOQUIAL NICKNAMES:
European Shore Crab

SCIENTIFIC NAME:
Carcinus maenas

Distinguishing Features and Behaviors

The green crab is one of the most common intertidal crabs found in New England. Green crabs have four pairs of legs which they use to scurry sideways. Their front pincers are almost equal, but one claw is slightly larger and blunt.

Green crabs are distinguishable from other intertidal crabs by their color and the shape of their carapace, although, when young, they are often mistaken for the white-fingered mud crab. Green crabs are shaped like a fan, but their carapace is usually square. Between the eye sockets are three sharp points or teeth, and five points run along the side of the carapace, curved toward the side of each eye socket.

These crabs are predators and scavengers, feeding mostly on and around mussel beds. Green crabs also prey upon small worms, mollusks and crustaceans. In turn they are a favorite food for gulls, herons and bottom fish such as the tautog.

Adult crabs forage in the subtidal shore following the tide and stay submerged much of the time. During the winter months, adults and large juveniles migrate into deeper waters of the Bay. Other juveniles remain in the harsh intertidal zone year-round and are frequently exposed with the receding tide.

The green crab is voracious, often called the "angry crab." It is an aggressive fighter and moves quickly, which helps it escape many confrontations. The green crab can tolerate a wide range of environmental extremes in intertidal zones, including low salinity levels, cold temperatures and drying out. It can withstand brackish conditions and is able to live in salinities as low as six parts per thousand.

Relationship to People

Introduced from Europe, green crabs have become one of the most common crabs along New England shores, including Narragansett Bay. Until a few decades ago, they were uncommon in Maine, but now they thrive there in the coastal rocky regions.

Green crabs are major predators on soft-shelled clams and are believed to be destructive to the population. They are also used extensively as bait, particularly in the recreational tautog fishery.

FIELD MARKINGS:
Dark to light green with yellowish mottling. Females have an orange-green back and a red abdomen. Size: up to 2.5 inches, head to tail. Carapace is 3 inches wide.

HABITAT:
Rocky shores and jetties, mud banks, salt marshes and tide pools.

SEASONAL APPEARANCE:
Year-round; moves to deeper waters in winter.

SENSITIVITY LEVEL:

Hermit Crab

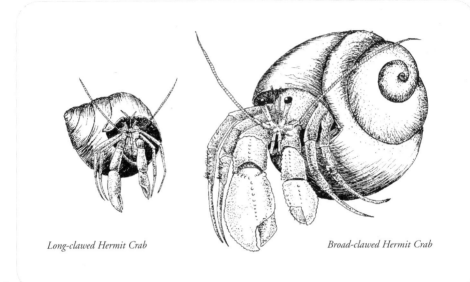

Long-clawed Hermit Crab

Broad-clawed Hermit Crab

COLLOQUIAL NICKNAMES:
Long-clawed Hermit Crab,
Broad-clawed Hermit Crab

SCIENTIFIC NAME:
Pagurus longicarpus,
Pagurus pollicaris

Distinguishing Features and Behaviors

Hermit crabs earned their name because they live in shells created by mollusks and have a tendency to withdraw into those shells when threatened. Two of the most common species of hermit crabs in Narragansett Bay are the broad-clawed and the long-clawed hermit crab, distinguished from each other by their size and the shape of their claws.

Hermit crabs are not considered to be true crabs because their exoskeleton does not cover the entire body. To protect their soft abdomen, hermit crabs steal shells formed by other mollusks — usually snails, periwinkles or whelks. Hermit crab bodies are curved and the last pair of appendages on the abdomen is modified to form a clamp, enabling the crab to maintain its position in the shell. It is very difficult to pull hermit crabs from their shells without injuring or killing them.

Long-clawed hermit crabs are the smaller species, inhabiting vacant periwinkle and oyster drill shells. They live mostly in shallow water and are common in tide pools and salt marshes. The major claw is narrow and usually hairless. The broad-clawed hermit crab — the largest hermit crab in Narragansett Bay — has broad, flat major claws and resides in large whelk or moon snail shells. They live in deeper waters of the Bay.

Hermit crabs are nomadic, always searching for new shells when they grow too big for their old ones. They move about actively, pausing only for the inspection of possible food or a new shell. A hermit crab will never leave the safety of its old shell until it finds a replacement. Larger crabs will often evict smaller crabs from a choice shell by using aggressive force.

As omnivorous scavengers, they eat small bits of fish, shore shrimp, dead plants, algae and other hermit crabs.

Relationship to People

Hermit crabs play an important role in the benthic community. As scavengers, they recycle nutrients back into the ecosystem. The meat found in the tails of larger hermit crabs is often used as bait in recreational fishing.

FIELD MARKINGS:
Long-clawed hermit
crabs have reddish-
tan bodies; claws are
white with a gray or
brown median
stripe. Size: up to
1.5 inches long.

Broad-clawed hermit
crabs are reddish,
gray or tan. Size:
up to 4 inches long.

HABITAT:
Rocky tidal zones,
tide pools, salt
marshes, open shores.

SEASONAL APPEARANCE:
Year-round.

SENSITIVITY LEVEL:

Lady Crab

COLLOQUIAL NICKNAMES:
Calico Crab, Sand Crab

SCIENTIFIC NAME:
Ovalipes ocellatus

Distinguishing Features and Behaviors

The lady crab, or calico crab, is a brightly colored, aggressive swimming crab. In the water and under direct sunlight, this crab's coloring appears iridescent. The species is called the lady crab because of the beautiful color patterns on the shell, although there are male lady crabs as well as female. The sharp, powerful pincers are whitish in color with purple-spotted tips and jagged teeth. The last pair of legs are modified into paddles, adapted to help the lady crab swim through the water. Three sharp points are present between the eye sockets, as well as five sharp points along the carapace that turn toward the eye sockets. The number of points along the carapace helps to distinguish this crab from the similar-looking blue crab.

Lady crabs are known for their aggressive disposition and sharp claws. This crab is often seen partially buried in sand with only its eyestalks protruding. When it sees prey, the lady crab will dart out of its hiding place using its powerful paddles to swim after prey. Like most other crabs, lady crabs are scavengers, eating both dead and live fish, crabs and other invertebrates. They can rapidly consume seed clams and prey upon hard clams. Lady crabs are preyed upon by oyster toadfish, tautog, striped bass, lobsters and other crabs.

The tail of the lady crab is tucked underneath the body and lies against the abdomen. On the female, it is shaped like a rounded triangle; on the male it is pointed and narrow. Female crabs use the tail to cover their eggs. Young crabs hatch in the early summer months, beginning their lives as zooplankton. They settle to the bottom by early fall.

Relationship to People

Lady crabs are migratory within Narragansett Bay, and, at certain times of the year, make up a large portion of the overall crab population. The lady crab will come close to the shore with the tide and is most often the crab that pinches bathers' toes at the beach.

The meat of lady crabs is not considered as tasty as that of other Bay crabs, and therefore, they are not harvested commercially.

FIELD MARKINGS:
White to yellow-gray, with reddish-purple mottled spots over entire body and claws. Size: 2 to 4 inches wide, 1 to 2 inches from head to tail.

HABITAT:
Sandy or muddy bottoms, often in shallow waters.

SEASONAL APPEARANCE:
Year-round.

SENSITIVITY LEVEL:

Mantis Shrimp

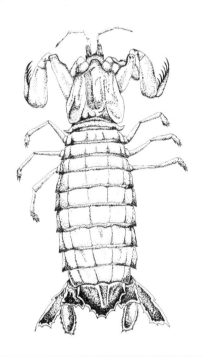

Colloquial Nicknames:
Shrimp Snapper

Scientific Name:
Squilla empusa

Distinguishing Features and Behaviors

One of the more elusive creatures is the mantis shrimp, which lives in burrows dug into the muddy bottom of the Bay. Its arm structure is similar to the praying mantis, with jackknife claws for forearms. The last part of the claw has five or six sharp spines that fold back into the claw, like the blade of a knife. The body of the mantis shrimp has sharp spines along the edges of the larger segments and three pairs of walking appendages.

The tail resembles that of a lobster and has a blunt ridge down the middle. The fanlike gills along the ventral abdomen serve to absorb oxygen as well as to ventilate their burrows. Retractable spines on the tail serve to anchor the mantis shrimp in its burrow. Its eyes are distinctively bright green and stalked above the head, providing the shrimp with almost 360-degree visibility.

Mantis shrimp are nocturnal, feeding on snails, shrimp, crabs and fish. They hunt by lying motionless in their burrows until they see desirable prey, then they lunge out and strike with their sharp claws. The claws of the mantis shrimp have been well studied and can break a pane of glass with their force.

Relationship to People

These shrimp are considered something of a nuisance by commercial fishermen because one quick snip with their claw can cut a shrimp or fish in two. They are edible and delicious but difficult to catch because mantis shrimp are nocturnal and live alone in burrows widely spaced apart. These burrows can be quite complex, with many exits and entrances.

Mantis shrimp have nasty temperaments, earning the nickname "thumb splitters" from fishermen who have been injured trying to remove them from their nets.

These creatures require high levels of dissolved oxygen in the water and will perish quickly in stagnant water of low-oxygen environments.

FIELD MARKINGS:
Body is grayish-blue with dark green or blue margins and bright emerald green eyes. Size: 8 to 10 inches long, 2.5 inches wide.

HABITAT:
Many-chambered burrows in the sand or mud of shallow and deeper waters.

SEASONAL APPEARANCE:
Spring through fall.

SENSITIVITY LEVEL:

Mole Crab

COLLOQUIAL NICKNAMES:
Beach Flea, Sand Crab

SCIENTIFIC NAME:
Emerita talpoida

Distinguishing Features and Behaviors

The mole crab is an egg-shaped crustacean with a smooth, convex carapace. An atypical crab shape, this crab's abdomen is broad in the front and tapers to its tail, which has a pair of forked, leaflike appendages. A long, spearlike tailpiece folds under the body, and a pair of dark eyes sits atop long, thin eyestalks.

Unlike most crabs and lobsters, the mole crab is somewhat defenseless, having no appendages to bait, sting or threaten predators. Instead, its

84

appendages are better adapted for digging. Its Latin name *talpoida* means "molelike," an appropriate name to describe its excellent digging abilities.

The mole crab has two highly functional pairs of antennae, both covered with fine, featherlike hairs. One pair is used for breathing, the other for feeding. Mole crabs bury themselves in the sand, anchoring their bodies in place with their tail plates and extending their antennae. The breathing antennae form a small funnel that takes in oxygenated water and filters sand grains away from its gills. The feeding antennae unfold into the water, trapping phytoplankton in their fine hairs. When its hairs are full of food, the mole crab ingests the phytoplankton by wiping the antennae across its mouth.

Mole crabs inhabit the shoreline where breaking waves crash, often leaving them exposed. When this happens, the crabs dig furiously into the sand to seek cover, escaping the next wave. The crabs adapt to the tidal cycle, migrating up and down with the high and low tides, always burrowing where the waves break. They are free swimming in the winter when both sexes move into deeper waters.

Relationship to People

Mole crabs are favored bait for striped bass fishing. They are easy to capture where the waves break along the shoreline by digging a few inches into the sand in the spot where the crab was last seen burying itself.

FIELD MARKINGS:
Pale, grayish-tan or sand-colored. Size: 3 inches long, 1 inch wide. The female is larger than the male.

HABITAT:
Open sandy beaches, in the surf zone.

SEASONAL APPEARANCE:
Year-round.

SENSITIVITY LEVEL:

Northern Lobster

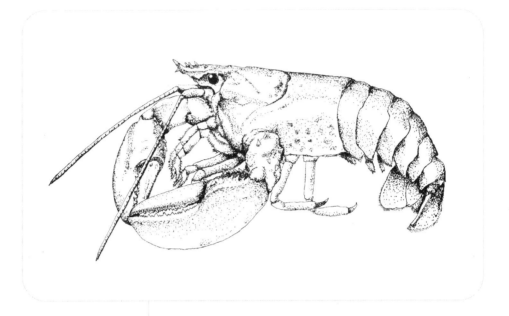

COLLOQUIAL NICKNAMES:
American Lobster,
Maine Lobster

SCIENTIFIC NAME:
Homarus americanus

Distinguishing Features and Behaviors

The northern lobster is a classic symbol of Narragansett Bay and the New England coast. Unlike the spiny lobsters found in tropical or sub-tropical waters, these creatures have a relatively smooth shell. The northern lobster is also distinguished by two large, specialized claws used for feeding and defense. A legal-sized lobster is usually about one pound and seven years old, but they can grow to sizes in excess of 20 pounds and may live for more than 50 years.

The underside of the tail is lined with swimmerets, which fan water around the burrow and help oxygenate water going to the gills. Although lobsters walk rather than swim, they can quickly escape a predator with a flap of the tail, scuttling backward with a quick burst of speed.

The sex of a lobster is determined by examining the first pair of swimmerets behind the last pair of walking legs. In females, this pair of swimmerets is soft and feathery, while in the males they are larger and more rigid. The tail of the female is much broader for carrying eggs.

Some lobsters are born with genetic defects causing them to produce extra proteins, which create atypical color patterns. This can cause the lobster to appear blue, yellow, albino or even split in color, an anomaly known as "half and half."

Lobsters are scavengers as well as predators, eating just about anything they can find along the bottom of the Bay, including fish, small crustaceans and mollusks. They are cannibalistic and have been known to devour other lobsters when in close quarters or in lobster traps. Northern lobsters are aggressive when fighting over territories and can inflict a painful pinch if disturbed. They are nocturnal and have highly developed sense organs on their legs to detect food in the absence of light.

Relationship to People

In the 1800s, lobsters were one of the most abundant species of Bay crustaceans. They were so plentiful that they could be captured with bare hands along the rocky shore at low tide.

Today, lobsters are considered a delicacy and are the most valuable single-species fishery in the Rhode Island economy. Lobster populations have declined in recent years due to disease, increasing water temperatures, degradation of habitat, overfishing and pollution.

FIELD MARKINGS:
Dark reddish-brown to olive green along the carapace, with lighter orange and white highlights and red and blue accents on the joints of the legs and claws. Size: averages 10 to 12 inches long from head to tail. Older lobsters may exceed 24 inches.

HABITAT:
Along rocky ledges in shallow waters of the Bay. Larger lobsters are found in deeper offshore waters.

SEASONAL APPEARANCE:
More common in summer; migrate to deeper waters in late fall.

SENSITIVITY LEVEL:

Northern Rock Barnacle

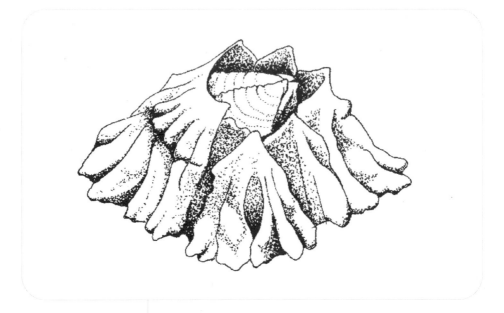

COLLOQUIAL NICKNAMES:
Barnacle, Acorn Barnacle

SCIENTIFIC NAME:
Semibalanus balanoides

Distinguishing Features and Behaviors

Upon first glance, there does not appear to be much more to a barnacle than the outer shell, but they are actually complex organisms that permanently attach their heads to a substrate with a kind of self-produced cement.

Barnacles have flat, irregular tops and resemble little volcanoes. The outer shell of a barnacle consists of white overlapping shelllike plates that grow with the organism. A soft cuticle covers the inside of the shell. The barnacle closes its shell when the tide goes out and opens it again when the tide comes in.

When the barnacle opens its mouth, featherlike feeding legs emerge to gather plankton from the water. These feathery legs are also used as gills to extract oxygen from the water. The animal grows by shedding its exoskeleton through molting. Molted barnacle exoskeletons can be seen floating on the water's surface.

Barnacles, like crabs and lobsters, are considered crustaceans because their feathery feeding appendages are jointed and they have an exoskeleton. They are the only crustaceans that remain fixed in one spot during their adult phase (as larvae, they are free floating).

Barnacles begin life as larvae in the planktonic column, often aggregating in large clusters when they settle to the bottom. Reproduction is sexual and requires the barnacles to settle in close proximity to one another. Barnacles will often attach themselves to slow-moving organisms, such as horseshoe crabs, lobsters and even whales. They are sensitive to extreme temperature and can die if the tide leaves them high and dry for too long.

Relationship to People

The barnacle's hard outer shell is very sharp and can cause cuts or scrapes on bare feet and legs. Barnacles are considered a nuisance by boaters, since they often attach themselves to boat bottoms, mooring lines, lobster and fish pots and pilings. The accumulation of barnacles on boats is called fouling and can reduce the speed of a boat and increase fuel consumption by 20% or more. Antifouling bottom paint is a toxic paint used to prevent the attachment of these organisms. Certain forms of this paint, although effective, can be harmful to other Bay life if not applied and removed properly.

The cement barnacles use to attach themselves to a hard substrate is currently being studied for dental applications.

FIELD MARKINGS:
Shell is white to gray, interior is darker.
Size: 1.5 inches in diameter.

HABITAT:
Permanently attached to rocks, pilings or any hard substrate in intertidal and subtidal habitats.

SEASONAL APPEARANCE:
Year-round.

SENSITIVITY LEVEL:

Rock Crab

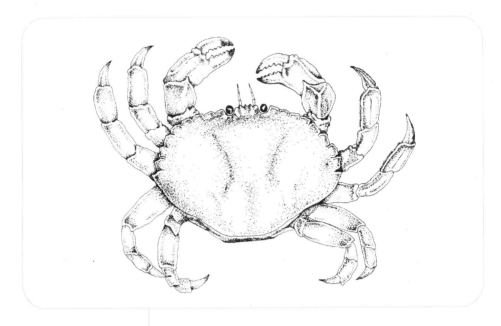

Distinguishing Features and Behaviors

Rock crabs are among the most common subtidal crabs in Narragansett Bay. They possess a hard shell, which enables them to live successfully in the harsh rocky tidal environment. Rock crabs have a relatively smooth oval or fan-shaped carapace with a rounded front border. Antennae, used for taste and smell, and two movable eyestalks are located at the front of the shell. Between the eyestalks are three spines, and nine smooth spines run along the outside edge of the carapace.

A similar crab is the Jonah crab *(Cancer borealis)*, which differs only slightly from the rock crab by having jagged spines along the shell and being more common in deeper waters.

The rock crab has two short front claws that are quite powerful but heavy and slow. The rock crab is a crawling crab and tends to move very little. Their short walking legs are covered in hairlike structures that function as sensory organs. Like all other crabs and lobsters, the rock crab grows larger by shedding its exoskeleton.

During mating season, the female releases a hormone into the water to attract a male. Mating can occur only during molting, with the male providing protection by encircling the female with his claws while she is soft-shelled and defenseless. In two or three days, once the female's shell has hardened, the male releases her. Juvenile rock crabs can be found in shallow, brackish, intertidal zones, while adults prefer deeper, saltier waters.

Rock crabs are eaten by fish, crabs, gulls and people. They are scavengers, and their primary prey includes worms, clams, mussels, other crabs and many other invertebrates.

Relationship to People
Rock crabs are considered excellent seafood and are harvested in the Bay both commercially and recreationally. Their population is large enough to support a commercial fishery from the Chesapeake Bay region north to the Canadian provinces.

FIELD MARKINGS:
Shell is yellow to red-orange with darker red mottling on top. The underside is whitish to a cream yellow. Size: up to 5 inches wide and 3.5 inches long at maturity.

HABITAT:
Rocky marine environments, jetties and tide pools, under and around rocks.

SEASONAL APPEARANCE:
Year-round.

SENSITIVITY LEVEL:

Sand Shrimp

COLLOQUIAL NICKNAMES:
Seven-spined Bay Shrimp

SCIENTIFIC NAME:
Crangon septemspinosa

Distinguishing Features and Behaviors

The sand shrimp is a small species of shrimp common to estuaries along the Atlantic coast. It has a stout, heavy body that tapers to a narrow tail. Similar in appearance to the shore shrimp, sand shrimp can be identified by a few key characteristics. The sand shrimp is flattened from top to bottom, rather than from side to side. Sand shrimp have only one pair of claws, which are shaped more like hooks than typical snapping claws.

The rostrum, a spiny projection between they eyes of many shrimp, is short on this species.

These shrimp can be observed in the near-shore salt marsh communities living in habitat similar to that of the shore shrimp. Most common in shallow water in summer months, this shrimp is inactive during the day, burrowing in the sediment with only antennae exposed. They remain burrowed throughout daylight hours but will emerge if the sediment is disturbed. At night they are active in the benthic community, foraging for food. Sand shrimp migrate into deeper waters in the winter, becoming inactive as the water gets colder. In the spring they will migrate back into shallow, warm estuarine waters.

Sand shrimp feed on benthic invertebrates, organic detritus and even larval and juvenile fish. They are preyed upon by bottom-dwelling fish, comb jellies and skates, and they are subject to cannibalism by their own species.

Relationship to People

Sand shrimp are one of the dominant predators in the benthic community. Primarily an estuarine species, they are able to withstand a wide range of temperatures and salinities but cannot exist in low oxygen conditions. Sand shrimp cannot tolerate freezing water temperatures, but in warm winters these shrimp eat the eggs and larvae of winter flounder, possibly having a major impact on that species.

Warming water temperatures cause the shrimp's eggs to hatch slightly earlier, which prolongs its reproductive cycle and life span.

FIELD MARKINGS:
Almost transparent pale to ash gray, spotted with brown or black. Size: up to 2.75 inches long.

HABITAT:
Among submerged seaweeds on sandy bottoms, salt marshes and eelgrass beds.

SEASONAL APPEARANCE:
Year-round; move to deeper water in winter.

SENSITIVITY LEVEL:

Shore Shrimp

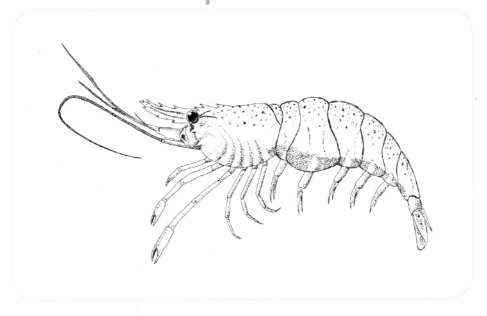

Distinguishing Features and Behaviors

Shore shrimp are the most common species of shrimp inhabiting New England's shallow coastal waters from Cape Cod south. They are found in salt marshes and seaweed along the Bay and in brackish tributary rivers.

Using well-developed sense organs, shore shrimp can easily maneuver and swim in the water, but they are found most frequently crawling along the bottom. Like other crustaceans, shore shrimp can cast off legs and regenerate new ones. They grow by molting — shedding their exoskeletons and forming new, larger coverings.

Shore shrimp are omnivores and feed on a range of plants and animals, including detritus, plankton and other small invertebrates. Between molts, a shore shrimp will eat almost anything, including its own exoskeleton.

Shrimp generally have thinner shells than their crab and lobster relatives. Their slender, elongated bodies are divided into three regions: the head, the thorax and the abdomen. The gills of the shrimp are located under the carapace and are oxygenated by a special organ near the mouth of the shrimp that pumps water over the gills. The body of the female shore shrimp is longer than that of the male, but the two are usually about equal in height.

Shrimp burrow during the day and move up to the surface to feed at night. They are particularly vulnerable to predation at this time and are preyed upon by many fish and larger invertebrates in Narragansett Bay.

Relationship to People

The shore shrimp is pollution-tolerant and found throughout the estuary. Unlike related species, the shore shrimp is smaller and generally not harvested as a human food source.

Because these shrimp are so common and consume algae and sea grasses, they play an important role in the ecology of the estuary. Shore shrimp are a major food source for larger predators in the Bay, such as fish and crabs. By breaking down detritus into tiny particles suspended in the water column, they provide a rich food source for smaller organisms.

FIELD MARKINGS:
Transparent gray with red, yellow, white and blue spots visible on their backs. Size: 1 to 2 inches long, half an inch wide.

HABITAT:
Among submerged seaweeds on muddy-sandy bottoms, ditches, salt marshes.

SEASONAL APPEARANCE:
Year-round.

SENSITIVITY LEVEL:

Spider Crab

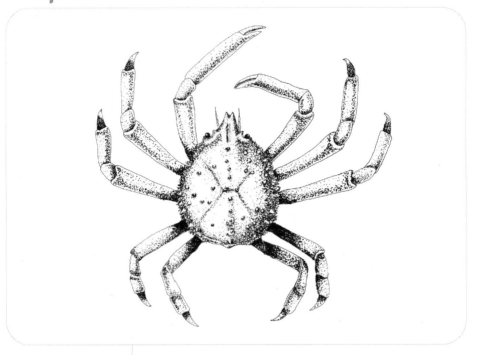

Distinguishing Features and Behaviors

The long-legged spider crab is one of the most widely recognized Bay crustaceans and is not a spider as its name suggests. The carapace of this crab is round and spiny, with nine small spines running down the center of the back. These crabs attach bits of algae, shell and seaweed to many fine, sticky hairs all over their bodies for camouflage.

The spider crab's tapered snout and short eyestalks are located on the rostrum, or tip, of the carapace, which extends out in a shallow, "V"-shaped notch. Spider crabs range in size, with adult males growing larger than juveniles and females.

The spider crab's claws are different from those of other Bay crabs. Their narrow, long pincers are slow and not as strong; however, the larger males have big claws that can deliver quite a pinch. The ends of the claws are used to scoop up bits of detritus and algae. The legs and pincers of the male spider crabs can be nearly twice as long as those of the female.

Spider crabs are non-threatening and somewhat lethargic scavengers. They have poor eyesight; however, they do have sensitive tasting and sensing organs on the end of each of their walking legs. This allows them to identify food as they walk over it. When startled, spider crabs will wave their pincers over their heads in a beckoning gesture to warn off potential predators.

Like all crabs, spider crabs molt to grow. The females stop molting after they become sexually mature and remain the same size for the rest of their lives. When molting, spider crabs will cling to the tops of eelgrass, close to the water's surface.

Relationship to People

One of Narragansett Bay's "living fossils," the spider crab's remarkable adaptability has enabled it to thrive practically unchanged for hundreds of millions of years. Spider crabs are highly tolerant of pollution and can live in harbors amid oil and other pollutants. They can also tolerate low oxygen, or eutrophic, environments where there are generally few inhabitants.

FIELD MARKINGS:
Body is mud-colored. Claws are whitish-yellow and stand out from the rest of the crab. Size: carapace is up to 4 inches wide. Males grow larger than females and can be 9 inches from claw to claw.

HABITAT:
Bay bottom, rocky shores, harbors, pilings.

SEASONAL APPEARANCE:
Year-round.

SENSITIVITY LEVEL:

American Oyster

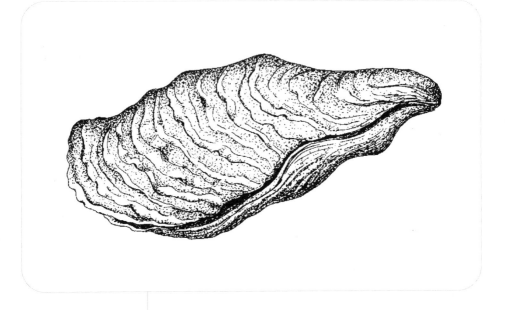

COLLOQUIAL NICKNAMES:
Common Oyster,
Eastern Oyster

SCIENTIFIC NAME:
Crassostrea virginica

Distinguishing Features and Behaviors

The American oyster is a bivalve mollusk, meaning it has two valves, or shells. The thick, lower valve is cemented to a hard substrate, and its shape depends upon the type of bottom to which it attaches. The upper valve is generally smaller and flatter. The hinge connecting the two valves is an elastic cushion held together by a thick, strong muscle. When feeding, this muscle relaxes, causing the shell to open and draw plankton-rich water into the mantle cavity. When the shell is closed, the mantle is filled with sea water, enabling the oyster to survive for extended periods out of water in cold or dry conditions.

Oysters provide a great volume of the hard substrate found in estuaries, often aggregating in clusters known as "oyster bars" or "oyster reefs." When they first settle out of the water column, juvenile oysters are known as spat. The oyster shells increase the area available to spat and other organisms that need a hard surface for settling. Oysters are preyed upon at various stages in their development by animals including oyster drills, sea stars, fish, crabs, flatworms, oyster catchers and mud worms.

Relationship to People

Oysters are among the best known, most valuable and most thoroughly studied of all mollusks. The native oyster population once thrived in the rivers and coves of Narragansett Bay, supporting a large oyster industry from the mid-1800s well into the 20th century. Native stocks were enhanced with spat imported from other estuaries like Chesapeake Bay. The industry was successful for many years but crashed in the 1950s as a result of increases in siltation, upper Bay pollution and the cost of buying oyster spat from other estuaries, and damage from the 1938 hurricane.

Rhode Island had some good oyster year classes in the early 1990s, but they have since dwindled in the Bay, probably due to disease. Save The Bay and other partners, particularly Roger Williams University, are working to restore oyster populations through aquaculture.

These mollusks play an important role in the Bay as filter feeders, removing bacteria and toxins from the water column. Based on the amount of toxins and bacteria found in their soft tissue, oysters are used as indicators of heavy pollution.

Oysters are famous for making pearls, which they do to get rid of sand or other irritants, although the pearls found in wild Bay oysters are irregularly shaped and not of much value as jewelry.

FIELD MARKINGS:
The exterior of the shell is grayish-white. Interior is white with a purple muscle scar. Size: 2 to 6 inches long and 2 inches wide.

HABITAT:
Rocky, sandy, muddy or shell-strewn bottoms below the tide line.

SEASONAL APPEARANCE:
Year-round.

SENSITIVITY LEVEL:

Atlantic Surf Clam

COLLOQUIAL NICKNAMES:
Surf Clam

SCIENTIFIC NAME:
Spisula solidissima

Distinguishing Features and Behaviors

The Atlantic surf clam is one of the largest species of clam inhabiting the Atlantic coast. Their heavy shells have a rounded, triangular shape, and the shell surface is smooth with fine concentric lines. New lines are added as the clam grows and can therefore be used to determine its age.

Surf clams have two short siphons which, when extended into the water column, are used to collect and expel water and plankton. The foot of the clam is used for movement. Surf clams burrow just below the surface of the mud and are often exposed at low tide, making them vulnerable to predation.

Surf clams are filter feeders, using one of their siphons to take in water and plankton. The other siphon is used to expel water after the plankton and oxygen have been removed. The gills extract oxygen and collect plankton on mucus-coated cilia. The cilia also send food particles to the mouth. The clam holds its shell tightly shut with two large muscles, called adductor muscles. These muscles are strong enough to prevent most predators from opening the shell.

Surf clams are preyed upon by gulls, who often drop them from the air onto hard surfaces such as boat decks, docks and parking lots in an effort to break open the shell. Other predators include whelks and sea stars.

Relationship to People

Surf clams make up about 70% of all clams commercially harvested in the United States. Only the adductor muscle of this clam is edible. They are used primarily in the production of canned clams and clam chowder and also as fish bait. These clams are harvested in offshore waters up to 100 feet deep. Empty shells are common along the southern beaches of Rhode Island and are prized among beachcombers.

During the 1996 North Cape oil spill off Moonstone Beach, hundreds of thousands of surf clams washed ashore in great masses, killed by the toxic oil.

FIELD MARKINGS:
Shell is yellowish-white to dark gray, with a brownish-black covering. Size: up to 6 inches long. Siphons can extend several inches outside the shell.

HABITAT:
Intertidally and subtidally, burrowed in sand or mud from the low tide line to 100 feet of water; beach shorelines.

SEASONAL APPEARANCE:
Year-round.

SENSITIVITY LEVEL:

Bay Scallop

COLLOQUIAL NICKNAMES:
Scallop

SCIENTIFIC NAME:
Argopecten irradians

Distinguishing Features and Behaviors

The bay scallop is one of the few filter-feeding bivalves that do not live buried in the sand or attached to rocks. Instead they settle and move freely along the bottom sediment surface. Unlike most bivalves, which are oblong or oval, the corrugated shell of the bay scallop is almost perfectly circular. The scallop has a strong hinge muscle within the shell but does not have a foot for digging or a siphon for water intake.

Along the edge, or mantle, of bay scallop shells are 30 to 40 bright blue eyes. Each eye has a lens, retina, cornea and optic nerve, enabling it to see movements or shadows and detect predators. Along the edge of the mantle are tentacles containing cells sensitive to chemical information in the water. These cells help the scallop react to its environment. When a scallop senses danger, such as a predatory sea star, it swims away by vigorously clapping the two sides of its shell together. This clapping movement forcibly ejects water from the mantle cavity through an exit near the hinge, propelling the bay scallop in a bouncing motion.

Scallops reach maturity when they are a year old, and they spawn in the summer. They grow quickly, rarely living past two years of age. When bay scallops are young, they attach themselves to objects such as eelgrass by means of a byssal thread. This helps them avoid bottom-feeding predators. As scallops grow, they drop to the sediment surface in the vicinity of eelgrass beds and move on to tidal flats to feed at high tide.

Relationship to People

The adductor muscle of the bay scallop makes excellent food, and scallops are harvested in the Bay using dredges, nets and rakes. In the 1880s, populations were so healthy that Rhode Island was a top supplier of scallops to New York. In the 1930s, as a result of wasting disease, many of the eelgrass beds in Narragansett Bay began to die off, leading to a severe reduction in the bay scallop population.

Populations have continued to decline in the Bay in recent years. Previously, one of the richest areas for scalloping in Rhode Island was Greenwich Bay. Today, it has no significant scallop population as a result of eelgrass degradation and loss caused by shoreline development and pollution.

Efforts are under way by state environmental groups to restore the scallop population in areas where they were historically found in the Bay.

FIELD MARKINGS:
Exterior shell ranges from gray to yellow or reddish-brown. Interior is white, often purplish near the hinge. Size: up to 3 inches in diameter.

HABITAT:
Subtidal zone, eelgrass beds, sandy and muddy bottoms, bays and harbors.

SEASONAL APPEARANCE:
Year-round.

SENSITIVITY LEVEL:

Blue Mussel

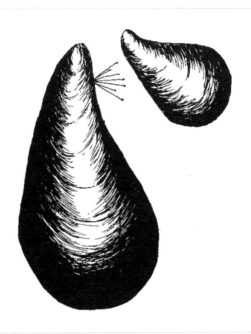

COLLOQUIAL NICKNAMES:
Edible Mussel

SCIENTIFIC NAME:
Mytilus edulis

Distinguishing Features and Behaviors

Shaped like a rounded triangle, the blue mussel is a hinged, filter-feeding bivalve. The blue mussel has a slender, brownish foot that allows it to temporarily hold onto a substrate, plus a strong, threadlike anchor, called a byssal thread, attaching it more securely to almost any substrate. The byssal threads are secreted as a liquid by a gland near the mussel's foot, and they harden upon contact with water. Byssal threads are tough but not necessarily permanent structures. To find protection or food, the blue mussel moves by releasing the byssal threads and using its foot.

Instead of the large protruding siphon common to hard-shell clams, the mussel has two short siphons on the inside of the shell, which direct the flow of water in and out. Mussels feed by filtering out detritus and plankton in the water. Cilia inside the mussel create a current that pulls in water and plankton.

Blue mussels live in dense colonies called mussel clumps. When the tide comes in, the mussel partially opens its shell and takes in water. Mussels resist dehydration during low tide by keeping their shells tightly closed. Their major predators are sea stars, whelks, fish, birds and humans.

The blue mussel — similar to another species, the ribbed mussel — is tough and withstands great temperature extremes, including freezing, excessive heat and drought. If a mussel is left exposed to air when the tide goes out, it survives by passing air over its moist gills to breathe. Blue mussels prefer areas of high salinity, while ribbed mussels are more prevalent in marshes where the salinity has been diluted by fresh water. Like clams, mussels have growth rings, which show their age. Full maturation takes from one to five years.

Relationship to People

Unlike the ribbed mussel, blue mussels are edible and are regularly harvested in Narragansett Bay. Many mussel beds are transitory in the upper Bay. The most permanent and most harvested mussel beds are in the East Passage and middle Bay. Deeper subtidal populations are of special interest both for commercial harvest and for their inter-ference with quahog growth and harvest. Overfishing and siltation caused by shoreline development have reduced the number of healthy mussel beds in the Bay.

FIELD MARKINGS:
Blue-black to brown outside with shiny violet interior. Size: up to 4 inches long, 2 inches wide.

HABITAT:
Intertidal shallow water along the shoreline and throughout the Bay; attached to rocks, pilings and shells.

SEASONAL APPEARANCE:
Year-round.

SENSITIVITY LEVEL:

Common Octopus

Distinguishing Features and Behaviors

Octopuses are solitary and territorial animals. When in danger, the octopus quickly changes its color and tries to escape while releasing a cloud of ink to confuse the predator. They create dens as a safe place to hide, remaining there for most of the day and leaving only at dusk for hunting trips. If they do need to leave during the day for food, the duration of the trip is much shorter. Octopuses feed mostly on bivalves and crustaceans.

Since they are experts at camouflage, finding them can be very difficult. Most octopuses leave piles of shell and crab pieces — called midden — outside their dens, which acts as a signpost for finding the octopus and also gives us information about their diets.

The common octopus only lives for one to two years. When octopuses mate, the female lays about 100,000-500,000 tiny eggs per breeding cycle. These eggs are laid in shallow waters and attached to a substrate for up to four months, during which time the female cares for them, rarely leaving or feeding. After the eggs hatch, the female octopus usually dies and the young live as plankton for one to two months.

Octopuses are very smart animals. They make observations based on visual and chemical cues with a complex brain that solves problems through experience. Their great eyesight is close to human eyesight with the ability to focus by moving the lens. Still, they don't rely solely upon it because their many suction cups provide octopuses with a great sense of touch.

Relationship to People

The octopus is a highly studied creature, known to have three hearts and blue blood. Some scientists believe that *Octopus vulgaris* is actually a set of subspecies, but haven't yet found how to separate the differences.

Octopus vulgaris is the most commercially fished species in the octopus fisheries, with numbers ranging from 20,000-100,000 tons caught each year. They are fished mostly for human consumption and are collected in octopus pots made of plastic or PVC. No bait is needed since the octopuses are attracted to this device as a safe place to live.

FIELD MARKINGS:
Skin is slightly bumpy with special pigment cells that allow the octopus to camouflage with its surroundings. The octopus' color also reflects its mood: white for fear and red for anger. Size: can reach 2-3 feet in length from the tips of the arms; males may be slightly larger than females.

HABITAT:
Along the coast and continental shelf, between the surface and down to 500 feet. Make their dens in rocks, reefs, and grass beds.

SEASONAL APPEARANCE:
Year-round; moving inshore in early spring to spawn, then leaving coastal waters in fall.

SENSITIVITY LEVEL:

Common Periwinkle

COLLOQUIAL NICKNAMES:
Winkle

SCIENTIFIC NAME:
Littorina littorea

Distinguishing Features and Behaviors

The periwinkle, which is related to limpets, whelks and other marine snails, is the most common snail in the Bay. Periwinkles are protected by a single spiral shell that grows with their bodies. Without the shell, they resemble land slugs. Their body includes a fleshy foot, a short tail and two antennae on the head. The cream-colored foot of the periwinkle is divided into a right and left half, which the snail moves alternately as the muscle ripples forward. Periwinkles use their foot to hold securely onto rocks or sea grasses when waves pass over them. Their stalked tentacles are sensory organs that are used to see and taste. Common periwinkles are closely related to the marsh periwinkle (*Littorina irrorata*), more common in salt marshes.

These creatures are herbivores, using their filelike tongue to feed on diatoms and algae attached to intertidal rocks. The periwinkle breaks down its food by mixing it with mucus on the radula before bringing the food into its mouth, and it can live for many days without food or water by retaining moisture in their gills. Periwinkles are eaten by sea stars, whelks and some fish. The shells of dead periwinkles are often inhabited by hermit crabs.

During spawning, females release fertilized eggs at night and only during high tide to ensure the egg capsules are dispersed and not exposed to air. After about six days, the eggs hatch into a larval stage that floats in the water column for several weeks before transforming into tiny periwinkles that settle in the subtidal zone.

Periwinkles close themselves into their shells and excrete a sticky mucus that hardens, firmly attaching the animal to a rock or blade of seagrass. Periwinkles adapt to a variety of environmental conditions, including extreme heat and wind when the tide is low and severe wave action and submergence when the tide is high.

Relationship to People

Although easy to spot along the coast, periwinkles are not native to North America or Narragansett Bay. Periwinkles were introduced to Nova Scotia from western Europe in the 1800s. Before their introduction, it is believed that the Bay's rocky shores were covered with lush green algae, unlike the gray bare rocks we see today. Although periwinkles are small in size, the constant scraping action of the radula of so many individuals can eventually lead to rock erosion.

Considered a delicacy in Europe, periwinkles are edible after a light boiling in seawater. Regularly harvested in the Bay, their local popularity as a food source has grown in recent years.

FIELD MARKINGS:
Shell is dark in color, usually brown, black or gray. Size: 1 inch long and three-quarters of an inch wide.

HABITAT:
Intertidal zones, rocky shores, tide pools, pilings, rock jetties.

SEASONAL APPEARANCE:
Year-round.

SENSITIVITY LEVEL:

Common Slipper Shell

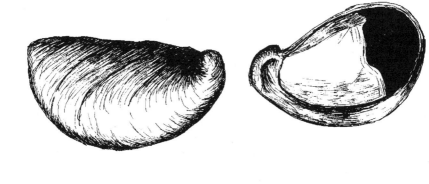

COLLOQUIAL NICKNAMES:
Slipper Shell, Lady's Slipper,
Slipper Snail

SCIENTIFIC NAME:
Crepidula fornicata

Distinguishing Features and Behaviors

One of the most common shells found along the Rhode Island coastline is the slipper shell. This shell is shaped like an egg or oval that has been cut in half, with the top of the shell turned sharply to one side. Looking at the underside, it is easy to see how the slipper shell got its name. Underneath the shell is a ledge to support the internal organs; the ledge extends about half the length of the animal.

Slipper shells are snails that aggregate, usually stacked on top of each other. They often form these aggregations when there is no hard substrate to which to cling. They attach to objects in large numbers and can sometimes suffocate the creature to which they are attached.

Different slipper shell species are characterized by different shell textures, including rough, smooth, ribbed, corrugated and flat. Although they do have a foot that they can use to move, by the time slipper shells reach maturity, they anchor themselves to a hard substrate and remain stationary.

Slipper shells use cilia to create water currents that flow through their mantle cavity. They are filter feeders, and as the water passes through the snail, it uses its radula to remove the food and bring it to its mouth.

These snails start their lives as males, but some change to females as they grow older. A waterborne hormone regulates the female characteristics. Once females, they remain females. Slipper shells often stack on top of each other for convenient reproduction: the larger females on the bottom, the smaller males on the top and the hermaphrodites in between. If the ratio of males to females gets too high, the male reproductive organs will degenerate and the animal will become female. The eggs are laid in thin-walled capsules, which the female broods under her foot.

Relationship to People

Slipper shells are a common beach shell on shorelines protected from heavy surf. Often found attached to moon snails, mussels, scallops and horseshoe crabs, they will also attach themselves to boats and dock pilings.

FIELD MARKINGS:
The exterior of the shell is dull white or cream with wavy longitudinal lines of a light chestnut color. The interior is a shiny light brown. Size: up to 1.5 inches wide, 2.5 inches long.

HABITAT:
Coastal marshes and inlets, tide pools, beaches.

SEASONAL APPEARANCE:
Year-round.

SENSITIVITY LEVEL:

Jingle Shell

COLLOQUIAL NICKNAMES:
Gold Shell, Toenail Shell

SCIENTIFIC NAME:
Anomia simplex

Distinguishing Features and Behaviors

The jingle shell is a bivalve mollusk (as are clams, oysters and scallops), meaning it has two distinct shells or valves. The upper valve is convex and movable while the fragile, lower valve is concave and matches the irregularities of the object to which it is attached. The lower valve remains in place through byssal threads, which are secreted by a gland near the jingle shell's foot and pass through a hole in the bottom valve to anchor the creature to the substrate.

Jingle shells are filter feeders, extracting plankton from the water they take in. Jingle shells lack a siphon; instead they have ciliated gills for food collection and storage. Despite a paper-thin, almost translucent shell, these creatures are tough. The shiny iridescence of the shell is retained even after death.

Because the shell of these animals is flat, round and often found in the company of oysters, jingle shells are commonly mistaken for juvenile oysters.

Relationship to People

Jingle shells are a popular shell for beachcombers, often used to make necklaces or wind chimes. It is named for the sound that is made when several shells are strung together to make a chime. The jingle sound can also be heard when waves beat down upon beaches covered with these shells.

Jingle shells are important commercially. They are strewn over oyster beds by shellfishermen in a process known as "shelling." Covering an area with jingle shells provides a good surface upon which juvenile oysters can settle.

FIELD MARKINGS:
Shell is shiny lemon yellow, golden-orange, brownish, silvery black or pale buff. Lower valve is white. Size: 1 to 3 inches in diameter.

HABITAT:
Shallow waters, beaches, oyster beds, mollusk shells.

SEASONAL APPEARANCE:
Year-round.

SENSITIVITY LEVEL:

Long-finned Squid

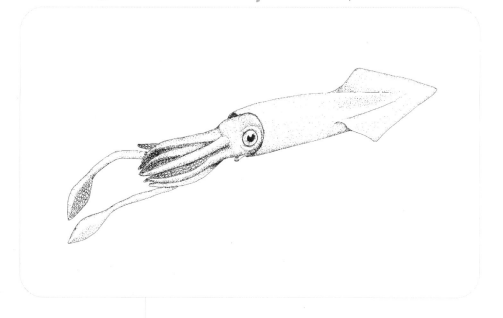

COLLOQUIAL NICKNAMES:
Squid

SCIENTIFIC NAME:
Loligo pealei

Distinguishing Features and Behaviors

Long-finned squid are the most common and abundant squid in Narragansett Bay and can be found here from spring through fall.

The long, flattened fins and aerodynamic, tube-shaped body of the long-finned squid help it to move gracefully and swiftly through the water. Squid move via a system of jet propulsion, filling their bodies with water and moving backwards with arms trailing behind. They can travel for long distances and are often found in schools.

Large, highly advanced eyes — similar in structure to the human eye — detect images and are used to locate prey. The squid has eight arms and two long tentacles with disk-shaped suckers used to grasp its prey. Rather than sticking like a suction cup, each sucker has a small barb that grasps onto the skin. The squid uses its two longest tentacles to capture the prey; the shorter arms help it hold onto the catch. The mouth of the squid looks like a bird's beak. The arms hold the fish near its mouth while the squid uses the beak to eat. Squid feed on fish, shrimp and other crustaceans.

Squid have speckles along their body that are special pigment cells known as chromatophores. When these cells enlarge, colored splotches are visible and can change according to the squid's behavior. This creature defends itself from predators in several ways. One method is to change into a warning coloration, pulsating purple. Another method is to squirt dark, purplish-black ink, which acts as a screen to blind predators such as sea bass, bluefish and mackerel.

Females lay gelatinous masses of eggs on the bottom; the eggs develop directly into embryos. Examining the egg masses under a microscope often reveals a developing and quite active embryo. The eggs aggregate in the water and are often mistaken for jellyfish. After spawning, the adult squid dies.

Relationship to People

Long-finned squid are fished commercially and used for bait as well as food. The popular Italian dish *calamari* is made from the arms, tentacles and body of the squid.

FIELD MARKINGS:
White with variable red, purple, yellow and brown speckles on its head and body. Size: 3 to 17 inches long, 2 to 4 inches wide.

HABITAT:
Deep water up to 300 feet.

SEASONAL APPEARANCE:
Offshore during late autumn and winter; return inshore early spring and summer.

SENSITIVITY LEVEL:

New England Dog Whelk

COLLOQUIAL NICKNAMES:
Dog Whelk

SCIENTIFIC NAME:
Nassarius trivittatus

Distinguishing Features and Behaviors

New England dog whelks are small, one-shelled mollusks distinguishable by their pointed spiral shells, raised beads along the ridges and sharp scalloped outer lips. Rarely solitary, this species of whelk is often found in clumps of many individuals.

Despite their small size, dog whelks are scavengers and predators. They feed mostly on dead fish but are capable of drilling holes into the hard shells of mollusks using a tooth-covered drill called the radula. After drilling the hole, the dog whelk feeds upon the soft flesh of the mollusk with its proboscis, an organ that includes the radula, mouth and esophagus.

Different methods of attack are implemented depending upon the prey. When feeding on barnacles, for example, the whelk will pry open the top plates rather than drilling a hole.

The color and shape of the dog whelk changes in relation to its geographical location and the prey it has consumed. Dog whelks that eat mussels are predominately dark reddish-brown, while whelks consuming barnacles are white to light yellowish-brown. Dog whelks living in high-energy environments, such as beaches, have smaller ridges than whelks from quieter marshes and inlets.

Relationship to People

The New England dog whelk is one of the most common snail species found along the Narragansett Bay shoreline, but it is well-camouflaged and moves slowly, making it inconspicuous.

Egg masses, resembling clumps of puffed rice, are commonly found on beaches in early summer.

FIELD MARKINGS:
The shell ranges in color from white to yellow-gray to dark reddish-brown, and occasionally has brown spiral stripes. Size: 1 inch long, half an inch wide.

HABITAT:
Rocky, sandy, shallow intertidal area.

SEASONAL APPEARANCE:
Year-round.

SENSITIVITY LEVEL:

Northern Moon Snail

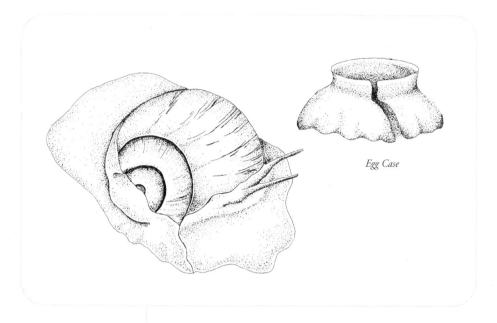

Egg Case

COLLOQUIAL NICKNAMES:
Sharks-eye Snail

SCIENTIFIC NAME:
Lunatia heros

Distinguishing Features and Behaviors

Moon snails are large, round, one-shelled mollusks that prey upon bivalve mollusks just below the surface of the sand. The large, fleshy foot of the moon snail is three times as long as the shell and enables it to move relatively fast, for a snail. They search under the sand for clams and other prey.

Moon snails have an organ called the proboscis, which includes the mouth, esophagus and radula, all of which are used for feeding. These creatures secrete an acidic material that softens the shell of their prey, allowing them to easily drill through it.

Once the hole has been made, the moon snail removes the flesh of its prey.

Moon snails are voracious yet selective predators, feeding solely on one species of bivalve in an area at a time, including other moon snails.

The sexes are separate, and males tend to be smaller than females. Mating occurs around the middle of summer, at which time the female constructs a collar-shaped egg case, often called a "clergyman's collar." The female secretes a gelatinous sheet from her shell in which the eggs are laid. Sand particles stick to the collar, making it tough and durable. Once detached from the shell, the collar is tough enough to withstand the elements until the eggs are ready to hatch. Egg collars can be found floating in shallow waters, along sandy bottoms and washed up on shorelines.

Relationship to People

Northern moon snails are edible but not widely harvested. A rapid population growth of moon snails can disrupt the balance of an ecosystem in an area of the Bay by decimating all the individuals in one area. Because they plow through so much sediment, moon snails will often destroy the tubes of mud-dwelling worms. Large groups of moon snails can consume entire areas of soft-shelled clams, destroying the bed and affecting commercial and recreational clam harvests.

FIELD MARKINGS:
Shell is light yellow to brown with dark streaks or whorls and a blue-gray foot. Size: shell is 2 to 3 inches long.

HABITAT:
Sandy, intertidal areas.

SEASONAL APPEARANCE:
Year-round.

SENSITIVITY LEVEL:

Oyster Drill

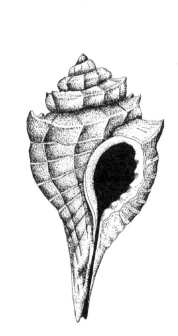

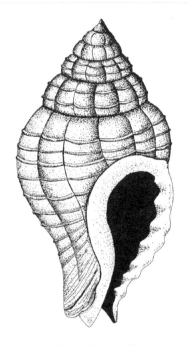

Thick-lipped Oyster Drill

Atlantic Oyster Drill

COLLOQUIAL NICKNAMES:
Thick-lipped Oyster Drill,
Atlantic Oyster Drill

SCIENTIFIC NAME:
Eupleura caudate,
Urosalpinx cinerea

Distinguishing Features and Behaviors

Oyster drills are small predatory snails. The Atlantic oyster drill and the thick-lipped oyster drill are the two species most common in Narragansett Bay. The shell of the Atlantic oyster drill is ribbed with a pointed spire, a flared outer lip with two to six teeth along the opening plus many rough and raised whorls and longitudinal ridges on the shell. The thick-lipped oyster drill is similar, except the anterior canal is longer with a smaller opening. This oyster drill resembles a small whelk or conch with distinct teeth.

The egg capsules of these snails can be found in spring and early summer. The eggs are small, urn-shaped, leathery cases generally found attached to pilings or mollusk shells.

Oyster drills are destructive snails that prey directly on small shellfish, most notably oysters. The oyster drill attacks its prey by making a small hole through the shell, using a drilllike organ called the radula. The radula is aided by the secretion of sulfuric acid to carve away the shell and make the hole. Once the hole is made, the animal will digest the soft meat of the prey. The mark left by an oyster drill can be identified as a wide, round hole tapering to a small point in the shell of a mollusk. This is similar to the hole made by the New England dog whelk, but different from many pock-mark holes made by boring sponges.

Relationship to People

Oyster drills are one of the worst enemies of small oysters, full-grown clams, oysters, other bivalves and barnacles. Oyster drills and sea stars can create serious problems for commercial and recreational shellfishing. When oyster drills occur in great numbers, they can decimate whole oyster and clam beds.

FIELD MARKINGS:
Dull brownish to gray shell with white points on both species Size: thick-lipped oyster drill grows up to three-quarters of an inch long. Atlantic oyster drill grows up to 1 inch long.

HABITAT:
Sea grasses, oyster beds, intertidal bottoms.

SEASONAL APPEARANCE:
Year-round.

SENSITIVITY LEVEL:

Quahog

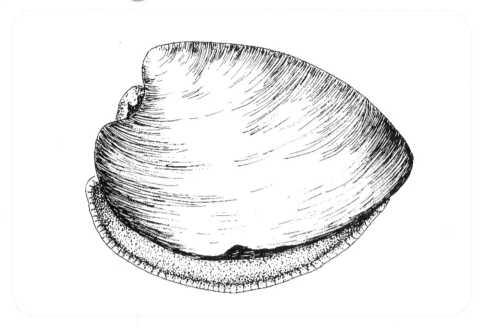

COLLOQUIAL NICKNAMES:
Hard-shell Clam,
Cherrystone, Littleneck

SCIENTIFIC NAME:
Mercenaria mercenaria

Distinguishing Features and Behaviors

Quahogs, or hard-shell clams, are bivalve shellfish that inhabit the mud flats of Narragansett Bay. The name quahog comes from the Indian name "poquauhock," meaning "horse fish."

Quahogs, which can live up to 40 years, are found along the temperate eastern seaboard from Canada to Florida. Their population is most concentrated in estuaries between Cape Cod and New Jersey where the salinity is less than that of the open ocean.

Quahogs do not remain fixed in one spot for life; they move through the mud using a muscular foot. With two short siphons, the quahog filters water in and out of its shell, absorbing plankton, bacteria and oxygen. Quahogs are extremely efficient filter feeders, and larger ones can filter about a gallon of water per hour.

Common predators include sea stars, whelks, crabs, snails, birds, some fish and humans. The entire body of the quahog is edible, not just the large adductor muscle that is found in larger species of clams. Empty shells with a small hole the size of a pencil point are evidence of consumption by moon snails, dog whelks or oyster drills.

Relationship to People

Quahogs are prized as a human food and constitute one of Narragansett Bay's most important fisheries. Unfortunately, human activity threatens the survival of this species. Since quahogs are filter feeders, they absorb bacteria and viruses from polluted waters; if eaten, these clams can make people sick. However, if pollution in the water stops, the quahog can clean itself simply through its regular filtering action. Shellfish are often used as an indicator of Bay health by measuring the pollutant levels of these animals in a specific area.

Narragansett Bay once provided 25% of the nation's supply of hard-shell clams. Due to pollution caused mainly by sewer overflows and stormwater runoff, about 60% of the Bay's shellfish beds are closed permanently or on a conditional basis. Conditional closures are determined by the amount of rainfall that occurs over the course of time.

The Latin name *Mercenaria mercenaria* is derived from a word that means "wages" and was given to the quahog due to the native peoples use of its purple inner shell or "wampum" as money and jewelry.

FIELD MARKINGS:
Shell color ranges from white to gray, with dark rings.
Size: 1 to 4 inches wide.

HABITAT:
Burrows just below the sand intertidally and subtidally.

SEASONAL APPEARANCE:
Year-round.

SENSITIVITY LEVEL:

Razor Clam

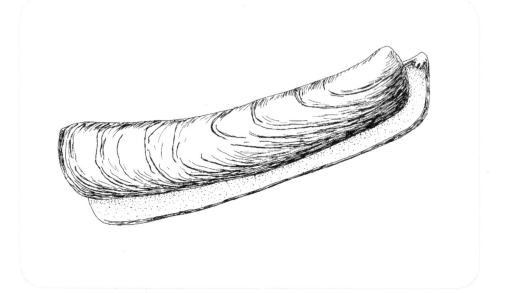

COLLOQUIAL NICKNAMES:
Straight Razor Clam,
Jackknife Clam

SCIENTIFIC NAME:
Ensis directus

Distinguishing Features and Behaviors

Razor clams are so named because their long, convex shape resembles an old-fashioned barber's straight-edged razor. The shell tends to be glossy with a purple region near the curving edge. The cream-colored muscular foot protrudes approximately five inches from the shell when extended. Like other bivalves, a siphon is present and the hole from which it protrudes is uniquely shaped, like a key hole.

This clam is well adapted for living in soft tidal substrates. Because of its short siphons, the razor clam burrows just below the surface to feed. When the tide goes out, it burrows quickly into deeper sand. The foot of the razor clam is larger and more agile than that of other bivalve clams. To burrow, the razor clam pushes its strong, narrow foot down deep into the substrate, then expands the foot as an anchor and pulls the body and shell deeper into the sand.

The razor clam is highly sensitive to vibrations and the impending danger of a nearby predator. Its strong muscular foot enables it to propel itself out of its burrow to escape an attack from below or to burrow quickly if a predator is attacking from above. Razor clams are preyed upon by clamworms and moon snails.

Relationship to People

Razor clams are edible but are not regularly harvested for consumption by people. They are extremely strong and almost impossible to remove from their substrate in one piece, as the shell can pull free from the body of the clam. Attempting to pull the razor clam from the burrow with bare hands can be dangerous, since the shell's edge is quite sharp. Clammers have found that pouring table salt into the burrow increases the salinity enough to irritate the clam, causing it to come to the surface.

FIELD MARKINGS:
Shell is brownish-green. Size: up to 10 inches, about six times longer than wide.

HABITAT:
Sandy and muddy bottoms of bays and estuaries, intertidal and subtidal zones.

SEASONAL APPEARANCE:
Year-round.

SENSITIVITY LEVEL:

Ribbed Mussel

SCIENTIFIC NAME:
Geukensia demissa

Distinguishing Features and Behaviors

Ribbed mussels are similar to blue mussels in shape and size but differ in appearance, color and habitat. The shell of the ribbed mussel is a long, rounded triangle with corrugated ribs along its length. Unlike the blue mussel that attaches to a piling or dock, ribbed mussels are usually found partially buried in the sediment.

When buried, two slitlike siphons with frilled edges extend from the body to the muddy surface. Gills lined with cilia aid the siphons in removing oxygen from the water and trapping plankton and organic matter. Particles of organic nutrients are processed into inorganic matter by the mussel and are then recycled back into the mud. This concentrated inorganic material helps to enrich the surrounding mud and contributes to salt marsh growth.

Ribbed mussels do not burrow completely into the muddy or sandy bottom but remain partially exposed. They anchor themselves with byssal threads, which are mucus strands attaching the mussel to the substrate. Burrowers such as soft-shelled clams must retract their siphons and close the shell tightly when the tide recedes to avoid desiccation, or drying out. Ribbed mussels, however, burrow in such a way that water is retained in the mantle when the tide recedes, an adaptation of great importance to life in the intertidal zone. They are able to withstand periods of drought and extreme fluctuations in temperature and salinity.

Ribbed mussels play a critical role in the health of a salt marsh by exhibiting a cooperative relationship with marsh plants and animals. Mussels establish habitat among the root structures of cordgrass and, in turn, provide essential nutrients that enhance the plant's growth. Mussel beds also provide support and stability for the root structures of cordgrass, allowing the plants to withstand harsh storm or ice conditions.

Relationship to People

Unlike their relative the blue mussel, ribbed mussels are rich in organic bacteria and are not commonly eaten by people.

FIELD MARKINGS:
Yellow-brown to
brownish-black on
top of the shell, with
glossy underside.
The body is lemon
yellow. Size: up to
4 inches long.

HABITAT:
Lodged among stems
and roots of cordgrass
in estuaries and salt
marshes.

SEASONAL APPEARANCE:
Year-round.

SENSITIVITY LEVEL:

Soft-shelled Clam

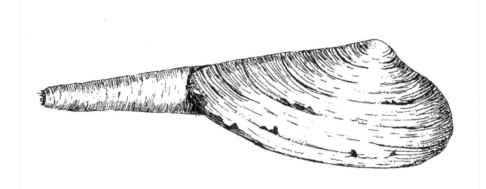

COLLOQUIAL NICKNAMES:
Steamer, Longneck

SCIENTIFIC NAME:
Mya arenaria

Distinguishing Features and Behaviors

Soft-shelled clams are thin, oval-shaped bivalves that can grow up to three inches long. They have two long siphons covered in a thick, black skin. Within the shell is a soft body composed of two gills, a heart, a stomach, a kidney and a large muscular foot. The gills are used for respiration and feeding.

The soft-shelled clam is a filter feeder. Water is brought into the clam through an intake siphon, which extends from its shell to the surface of the mud. As water passes over the gills, oxygen is removed for respiration and small hairlike structures, known as cilia, trap plankton for food. The particles of food are transported to the mouth, and water is expelled through the exit siphon. When the tide is high, the siphons extend out of the burrow; they are retracted during low tide. The siphons are encased in a fleshy tube and cannot be fully retracted into the shell. Because soft-shelled clams burrow, the siphons may be the only part of the clam a beachcomber will see.

Soft-shelled clams spawn in early summer. The eggs develop into free-swimming larvae, or plankton, and eventually settle onto a hard substrate, attaching themselves with a sticky thread secreted from a large, mobile foot. This byssal thread keeps them from being swept away by waves. Once mature, the foot reduces in size, and the clams release the substrate to burrow into the sediment where they remain for life.

The primary predator of the clam is the moon snail. Moon snails secrete an acidic material that softens the shell, allowing them to easily drill through and eat the clam. Large groups of moon snails can destroy soft-shelled clam beds. Soft-shelled clams are also preyed upon by digging fish such as the sea robin, which disrupts the burrows. Once dislodged, a soft-shelled clam cannot burrow quickly and is easy prey.

Soft-shelled clams can tolerate low salinities and rapid salinity changes. This allows them to thrive in upper as well as lower estuaries.

Relationship to People

Soft-shelled clams — known regionally as "steamers" — are commercially fished for use in clam chowder and are widely enjoyed steamed in water. Unfortunately, bacterial pollution has caused the permanent or conditional closure of many Narragansett Bay shellfish beds.

FIELD MARKINGS:
Shell is chalky white to dark gray. Size: up to 4 inches long. Siphons can extend several inches out of the shell.

HABITAT:
Burrowed in sandy or muddy bottoms of bays and estuaries.

SEASONAL APPEARANCE:
Year-round.

SENSITIVITY LEVEL:

Whelks

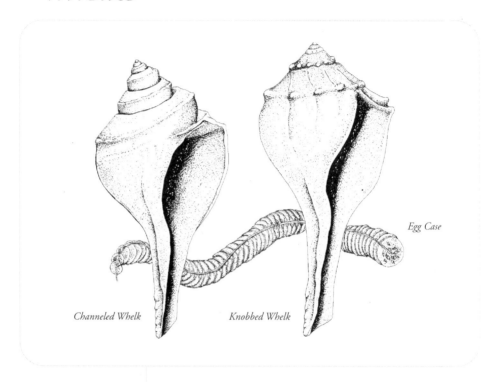

Channeled Whelk

Knobbed Whelk

Egg Case

COLLOQUIAL NICKNAMES:
Channeled Whelk,
Knobbed Whelk

SCIENTIFIC NAME:
Busycon canaliculatum,
Busycon carica

Distinguishing Features and Behaviors

Whelks are large snails with massive shells. The two most common species in Narragansett Bay are the knobbed whelk and the channeled whelk.

The knobbed whelk is the largest marine snail in the Bay. It is pear-shaped with a flared outer lip and knobs on the shoulder of its shell. The channeled whelk is generally smaller and has spiral lines instead of knobs deeply channeled on its shell.

Both species of whelk have an operculum — a hard, brownish-black, shelllike organ used to close the opening of the shell for protection. The feeding organ is called the proboscis, which includes the mouth, esophagus and radula. Whelks use the radula, a drilllike organ with small teeth, for grasping either flesh or plants. Whelks also have a long, tubular siphon, which they use to draw in oxygenated water.

Both whelk species are aggressive predators, preying on other invertebrates, particularly clams. They feed by prying a gap between the two valves of the clam and forcing the shell open with their strong muscular foot. As soon as the valves open, even the tiniest amount, the whelk wedges in the sharp edge of its shell, inserts the proboscis and devours the soft body of the clam.

Mating occurs by way of internal fertilization; sexes are separate. The egg casing of the whelk is a long strand of yellowish, parchment-like disks, resembling a necklace — its unique shape is sculpted by the whelk's foot. Egg cases can be two to three feet long and have 70 to 100 capsules, each of which can hold 20 to 100 eggs. Newly hatched channeled whelks escape from small holes at the top of each egg case with their shells already on. Egg cases are sometimes found along the Bay shoreline, washed up with high tide debris.

Relationship to People

Both channeled and knobbed whelks scavenge and hunt for food and are often found in crab traps and lobster pots, presumably stealing the bait.

Whelks are a favored food in the Northeast and are harvested all along the Atlantic coast. *Scungili*, a popular Italian dish, is made from the foot of both species of whelk.

FIELD MARKINGS:
The shell of both species is yellow-red or orange inside and pale gray outside. Size: channeled whelk grows up to 8 inches long; knobbed whelk grows up to 9 inches long and 4.5 inches wide.

HABITAT:
Sandy or muddy bottoms.

SEASONAL APPEARANCE:
Year-round.

SENSITIVITY LEVEL:

Common Sand Dollar

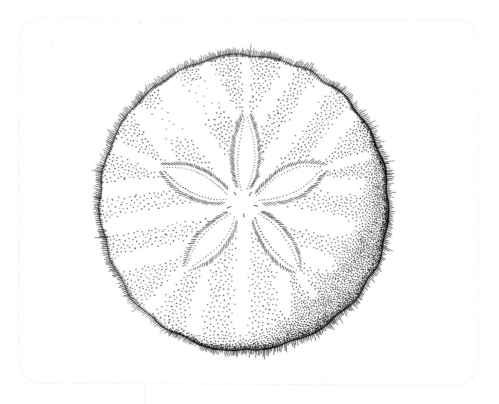

COLLOQUIAL NICKNAMES:
Sand Dollar

SCIENTIFIC NAME:
Echinarachnius parma

Distinguishing Features and Behaviors

The common sand dollar is circular with a maroon-colored velvet body. The shell has many small holes creating a petallike design, which is used to move water into its water-vascular system. Sand dollars can grow to about two to four inches in diameter.

Although they lack the five arms that most echinoderms possess, the sand dollar does have the five-part radial symmetry and is able to right itself if thrown on its back. Due to their small internal body and hard skeleton, they have very few predators .

The shell, or test, of the sand dollar has many small, brown spines that make it appear to be covered in velvet. These spines are not used for protection but help the sand dollar to crawl or burrow. It takes the common sand dollar about ten minutes to completely burrow into the sand, while other echinoderms take only a few minutes.

The spines of the sand dollar are also covered in small hairs and a slime coating, which, along with its tube feet, assist the animal in moving food to its mouth. The mouth — found in the center of the sand dollar's underside — has large, triangular teeth for feeding on algae and organic material.

Sand dollars are usually seen together in the intertidal zone where they burrow just under the sand. They can be found along the East coast from New Jersey north.

Relationship to People

Sand dollar eggs are commonly used in mitosis studies to help people understand cancer and other diseases.

Often while walking along a beach after a storm, people will find the skeletons of sand dollars washed ashore and bleached by the sun.

FIELD MARKINGS:
Circular with a maroon-colored velvet body. The shell has many small holes creating a petallike design. Size: grows to about 2-4 inches in diameter.

HABITAT:
Intertidal zone along sandy bottoms.

SEASONAL APPEARANCE:
Year-round.

SENSITIVITY LEVEL:

Purple Sea Urchin

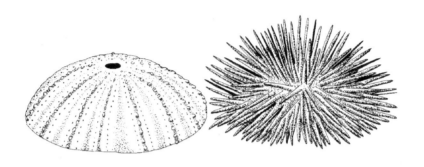

COLLOQUIAL NICKNAMES:
Sea Urchin

SCIENTIFIC NAME:
Arbacia punctulata

Distinguishing Features and Behaviors

Sea urchins are squat, round creatures with a hard exoskeleton usually covered in spines. They belong to the group of echinoderms that also includes sea stars and sand dollars. The hard exoskeleton, known as a test, is composed of calcium carbonate plates called ossicles. The exoskeleton can often be found among other shells on the shore, distinguishable by the sutures of the plates along with rows of bumps and pin-sized holes.

While alive, the exoskeleton resembles a pin cushion, covered in spines. Spines are absent in a small area near the top where pores lead to reproductive organs and on the bottom where the mouth is located.

In tidal areas, urchins use their spines to dig small depressions among the rocks and sand. These holes hold water even after the tide has gone out; the holes also protect the urchin from waves at high tide.

Urchins have an organ in their mouths called "Aristotle's Lantern," which they use for feeding. This structure resembles a bird beak, with five plates rather than two, and is used to scrape algae from rocks. Purple sea urchins are omnivorous and feed on sponges, algae, small invertebrates and detritus.

Relationship to People

Urchins are sensitive to light and hide in rock crevices during daylight; however, this has not prevented divers from finding and harvesting them. Sea urchins are harvested along both the Pacific and Atlantic coasts, but many areas have been closed in California and the Gulf of Maine as a result of overfishing. They are primarily exported to Japan for food.

Purple sea urchins — and the occasional green sea urchin — are common along the rocky shores of Narragansett Bay. Stepping on or touching the spines of an urchin can cause a small cut.

Field Markings:
Dark purple to reddish brown. Size: up to 2 inches in diameter. The spines can be up to 1 inch long.

Habitat:
Attached to rocks and shells in tide pools, on seaweeds, along rocky bottoms.

Seasonal Appearance:
Year-round.

Sensitivity Level:

Sea Cucumber

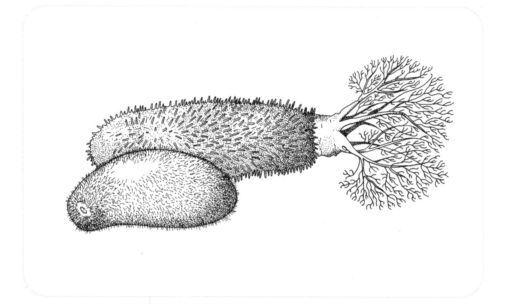

COLLOQUIAL NICKNAMES:
Hairy Sea Cucumber,
Common Sea Cucumber

SCIENTIFIC NAME:
Sclerodactyla briareus,
Thyone briarus

Distinguishing Features and Behaviors

Sea cucumbers are cylinder-shaped, nocturnal, brainless echinoderms with a muscular, leathery body lined with small tube feet. They have the five-part symmetry, but their millions of microscopic spines are found within their skin. Unlike other echinoderms, they are able to retain body fluid and breathe by pumping water in and out through an internal respiratory tree. Sea cucumbers have a short lifespan of five to ten years.

Sea cucumbers have an unusual way of deterring predators. When threatened, they can expel some of their body fluids to scare away their attacker. If that doesn't work, they eject most of their body organs to confuse the other animal. These organs can later be regenerated. Sea cucumbers feed mostly on plankton caught by the branched tentacles in front of their mouths. Their largest predators are crustaceans, fish and people.

There are more than 1100 species of sea cucumbers throughout the world, with the largest — the Tiger's tail — reaching six and a half feet in length.

The hairy sea cucumber's body is about four inches long and two inches wide, covered with tiny tube feet; the thickest part is the middle. This invertebrate has ten long tentacles surrounding the mouth, and they can be pulled into its body when startled. A beachcomber will usually see only the these tentacles emerging from a substrate. The sea cucumber's color ranges from black to brown to purple.

The hairy sea cucumber is usually found in shallow muddy waters, where the salinity is slightly high, along the East coast from Massachusetts to Florida. However, other species of sea cucumbers are found in a variety of habitats from warm waters to deep cold trenches.

Relationship to People

Many sea cucumbers ingest mud and sand to absorb organic matter. This type of feeding can act as a natural filter in the benthic community. There is a small commercial industry for sea cucumbers for human consumption.

Field Markings:
Black, brown or purple. Body is covered with tiny tube feet. Size: about 4 inches long and 2 inches wide.

Habitat:
Shallow muddy waters.

Seasonal Appearance:
Year-round.

Sensitivity Level:

Sea Star

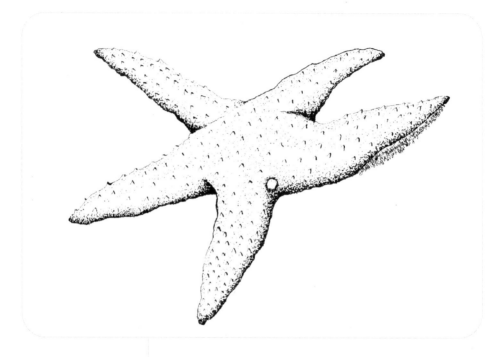

COLLOQUIAL NICKNAMES:
Forbes Star, Starfish

SCIENTIFIC NAME:
Asterias forbesi

Distinguishing Features and Behaviors

Sea stars are not fish as their nickname "starfish" suggests. They belong to a group of animals called echinoderms, which means "spiny skin," and are related to sea urchins, sea cucumbers and sand dollars.

Sea stars have five arms, or rays, connected to a small round body. Sea stars detect light with five purple eyespots at the end of each arm. The bright orange dot in center of the body is called the madreporite, and it pumps water into the sea star's body. This pumping action creates suction at the end of hundreds of tube feet, located in paired rows on the underside of the arms. Sea stars use this suction for movement and feeding. They wrap their bodies around hard-shelled clams and other bivalves, using the suction from their tube feet to pull shells apart. When the clam is opened, the sea star pushes its stomach out of its body and into the clam, secreting enzymes that digest the clam's soft body. The liquefied clam is then absorbed into the sea star's stomach. They feed often, and their size depends on the amount of food they eat, not on their age.

When low tides leave sea stars exposed, they are eaten by bottom-dwelling fish and crabs as well as seagulls. Similar to other Bay creatures like lobsters and crabs, sea stars can grow new arms if they lose them. This regeneration will occur as long as one-fifth of the sea star's body remains intact.

Sea stars breed in the spring, producing as many as 2,500,000 eggs. Females feel plump and spongy when their arms are filled with eggs.

Relationship to People

Sea stars pose a threat to commercial and recreational shellfishing efforts. One sea star can devour over 50 young clams in a week. When their population grows, sea stars can consume entire beds of shellfish. In 1929, an oyster company removed more than 10 million sea stars from 11,000 acres of oyster beds in Narragansett Bay.

Sea stars are harvested with a tool resembling a large mop that is dragged along the bottom. The sea stars attach themselves to the mop strands and are hauled aboard fishing boats. After removal, the sea stars are ground up and sold for fertilizer and poultry feed.

FIELD MARKINGS:
Brownish-purple to orange with lighter underside. Size: up to 12 inches across.

HABITAT:
Rocky shores, tide pools, dock pilings, Bay bottoms.

SEASONAL APPEARANCE:
Year-round.

SENSITIVITY LEVEL:

Short-spined Brittle Star

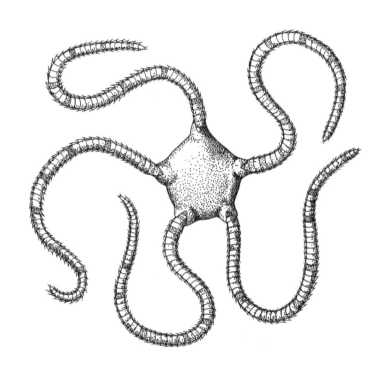

SCIENTIFIC NAME:
Ophioderma brevispina

Distinguishing Features and Behaviors

The short-spined brittle star is an echinoderm, usually green or brown in color. Its spine lies flat against its arms, with seven short spines on either side of each arm joint. The diameter of the central disc is about half an inch, with an arm length of one and a half to two and a half inches long; however, their size is dependent on their predation.

The word "brittle" means "easily breakable." These creatures are so named because they have the ability to shed a limb in an instant to avoid predation. Like their echinoderm relatives, they are also able to regenerate body parts.

Brittle stars do not rely on their tube feet for movement. Rather, they use their five long, whiplike arms, which are capable of fast, snakelike motions. These inverts are very flexible, highly mobile and easily camouflaged among plants. They can be seen at night in large groups crawling along the sea floor.

Brittle stars feed mostly on decaying material and plankton. They have no eyespots like other sea stars, but are able to sense some light though their body.

There are over 2,000 species of brittle stars found throughout the world. They are often found in tide pools, hiding in algae and under rocks and shells or, sometimes, disguised in dark, heavy algal mats from Cape Cod south.

Relationship to People
Scientists have found brittle stars to be sensitive to polarized light and the species may be used as an indicator of harmful levels of ultraviolet radiation.

FIELD MARKINGS:
Green or brown in color. Size: diameter of the central disc is about half an inch, with an arm length reaching 2.5 inches.

HABITAT:
Tide pools, under rocks or shells.

SEASONAL APPEARANCE:
Year-round.

SENSITIVITY LEVEL:

Bryozoans

Zooids (magnified)

Encrusting Bryozoan

COLLOQUIAL NICKNAMES:
Red Crust, Moss Animals

SCIENTIFIC NAME:
*Cryptosula
and related species*

Distinguishing Features and Behaviors

Bryozoans are animals that group together in large colonies of many individuals, giving the appearance of ground moss. In fact, their plantlike appearance earned bryozoans the nickname "moss animals."

The tiny individuals making up a bryozoan colony are called "zooids" and come in many different shapes, including oval, tubular, box and vaselike. Zooids are surrounded by a boxlike exoskeleton of secreted calcium carbonate or limestone. These exoskeletons range from slightly stiff to hard.

Common octopus, pg. 106

This green crab (pg. 76) exhibits a regenerated claw that hasn't yet achieved pigmentation.

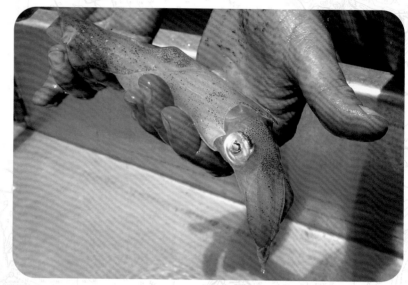

Long-finned squid, pg. 114

Narragansett Baykeeper John Torgan (right) with his father Philip exhibit their catch of the day, a bluefish (pg. 190).

Harp seal, pg. 292

Harbor seal, pg. 290

Northern sea robin, pg. 166

LEFT TO RIGHT: *Bayberry (pg. 42) and beach plum (pg. 43) are just two of the 20 native drought- and salt spray-tolerant plant species present in the Save The Bay Center coastal buffer.*

Mallard, pg. 244

LEFT TO RIGHT: A spider crab (pg. 96), blue mussels (pg. 104) and a purple sea urchin (pg. 134) cohabitate in a touch tank at the Save The Bay Exploration Center.

Short-spined brittle star, pg. 140

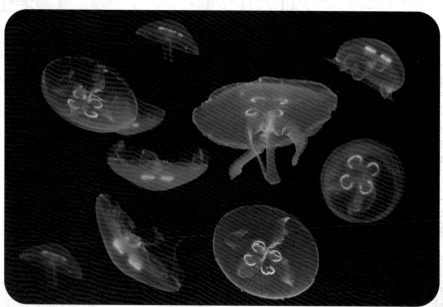

Floating moon jellies (pg. 56) on view at the Save The Bay Exploration Center.

Save The Bay works to restore the Bay's historic eelgrass beds (pg. 2), home to flounder, scallops and crabs. Photo by Tom Freeman.

A family of Canada geese (pg. 232) walk the Save The Bay Center grounds.

Oyster toadfish, pg. 168

Asian shore crab, pg. 66

Phragmites australis, pg. 26

Less than 100 pairs of osprey (pg. 282)
nest in Rhode Island, including this bird,
flying over Fields Point with dinner.

Sea star, pg. 138

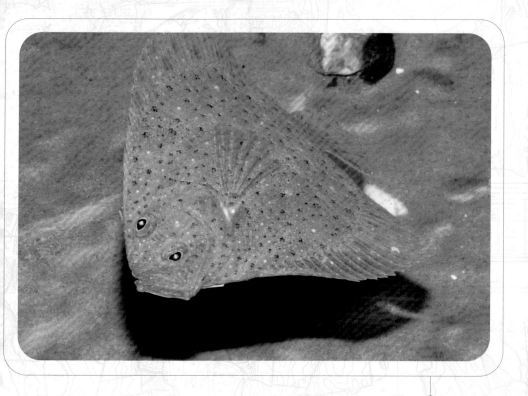

The summer flounder (pg. 196) is also known as a "left-eyed" flounder because its right eye migrates to the left side of its body.

Tautog, pg. 174

Virginia rose, pg. 45

Atlantic horseshoe crabs (pg. 68) mate in the salt marsh at the Save The Bay Center.

At left, the underside of the horseshoe crab displays its legs and "book gills."

The only part of the individual bryozoan that extends outside the exoskeleton is the lophophore, a spiral-shaped organ with ciliated tentacles used to strain plankton and detritus from the seawater. Some bryozoans seal themselves within their exoskeletons by shutting a flap similar to the operculum of a whelk to protect themselves.

There are 4,000 species of bryozoans, most of which live in salt water. Bryozoan colonies are generally classified either as bushy colonies or as calcareous encrustations; the latter is most common. Some encrusting species of bryozoans grow on the shells or exoskeletons of other invertebrates. Bryozoans also attach themselves to other hard objects in the water, including larger and more mobile animals.

Relationship to People

Along with clams and oysters, bryozoans eat bacteria filtered out of the water and serve as part of Narragansett Bay's natural pollution-filtration system.

Bryozoans are found on docks and other artificial structures and can be a nuisance when they attach themselves to the bottoms of boats and boat moorings. Accumulation of organisms on boats is called fouling and can reduce the vehicle's speed and efficiency. Antifouling bottom paint is commonly used to deter these animals from attaching but can be harmful to other Bay life if not used properly. Alternative antifouling products exist that can be effective without harming other animals and plants.

FIELD MARKINGS:
Colonies vary in color from light yellow to orange or brown and include some bright red and burgundy varieties. Size: individuals are one-fiftieth of an inch long.

HABITAT:
Intertidal region to great depths.

SEASONAL APPEARANCE:
Year-round.

SENSITIVITY LEVEL:

Sea Squirt

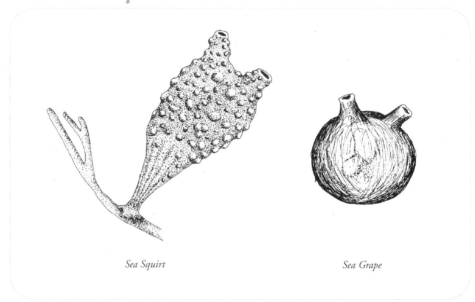

Sea Squirt   Sea Grape

COLLOQUIAL NICKNAMES:
Sea Grape

SCIENTIFIC NAME:
Molgula manhattensis

Distinguishing Features and Behaviors

Sea squirts — so called because they squirt water out of the siphons when touched or squeezed — are irregularly shaped creatures belonging to a group of animals called tunicates. Tunicates are classified as higher invertebrates because they are chordates — animals that have a spinal cord and a backbone during development. Other chordates include fish, birds, mammals, reptiles and amphibians.

Sea squirts can be found in three forms: as solitary animals, as a loose colony resembling a bunch of grapes and in a compound, gelatinous mass. The leathery tunic or body covering is composed of cellulose, a compound commonly found in plants.

Tunicates are filter feeders, constantly drawing water into their siphons, where planktonic particles are trapped in mucus-covered cilia that line the mouth and intestine. Water and waste products are excreted through an outgoing siphon. The round, grapelike tunicate has two openings at the top used to siphon water in and out of the body.

Tunicate larvae resemble tadpoles and live in the planktonic zone until settling. The free-swimming larvae eventually attach themselves to an object or other organism and will remain in the same place throughout their lives. Sea squirts will not settle unless the substrate is free of other organisms. They are usually found separate from other sessile animals.

Relationship to People

Sea squirts are tolerant of pollution and can survive in very dirty harbors.

Although some species of tunicates are native to Narragansett Bay, several of the more common species found here are actually invasive, often arriving in ship ballast water. Invasive tunicates can reproduce quickly, accumulate on boats and pilings, and smother organisms on the sea floor.

FIELD MARKINGS:
Golden yellow to purplish-brown, with white or purple specks.
Size: up to 1.375 inches high.

HABITAT:
Attached to the base of pilings, on rocks and buoys in intertidal zone.

SEASONAL APPEARANCE:
Year-round.

SENSITIVITY LEVEL:

Zooplankton

1: *Shrimp Larvae*
2: *Bivalve Larvae*
3, 4: *Barnacle Larvae*
5: *Crab Larvae*
6: *Copepod*
7: *Fish Larvae*
8: *Fish Egg*
9: *Shrimp Larvae*
10: *Comb Jelly*
11: *Copepod*
12: *Crab Larvae*

COLLOQUIAL NICKNAMES:
Microscopic animals,
Copepods

Distinguishing Features and Behaviors

Zooplankton is the common name given to many small species of animals found in fresh and marine waters throughout the world. The word "zooplankton," derived from Greek, means "wandering animals." Although some species can reach eight feet long, most of these animals are so minute they are visible only with a microscope.

Two general groups of zooplankton exist: holoplankton (those that remain planktonic throughout their entire life) and meroplankton (those that are larval stages of larger life forms). Jellyfish are the largest example of holoplankton. They remain in the pelagic zone for life and can grow as large as eight feet, with tentacles up to 120 feet. Meroplankton are the eggs and larvae of nearly all species of fish and benthic invertebrates. These creatures are planktonic during their developing stages and will eventually settle out of the planktonic zone as juveniles.

Of the numerous zooplankton species, the most abundant and diverse are copepods. Copepods are crustaceans similar to lobsters, crabs and shrimp. Their tough exoskeleton is composed of calcium carbonate, and their bodies are divided into three sections: head, thorax and abdomen. Two antennae protrude from the head and aid in swimming, while two to four pairs of appendages extend from the thorax.

Zooplankton migrate vertically in the water column each day, feeding on the phytoplankton near the surface of the water. They have adapted various mechanisms to float in the water column and protect themselves from predation. Some, such as larval crustaceans, have spikes that protect them and increase surface area for better flotation. Some species of fish larvae have oil globules that give them added buoyancy.

Zooplankton are a critical element in the Bay food chain, preyed upon by every filter-feeding organism, including shellfish, fish and whales. The great whales feed entirely on one particular zooplankton species called krill. Copepods and other zooplankton feed on phytoplankton and are the first link between the primary producers and larger animals. They are, by far, the most abundant group of animals in the world's oceans.

Relationship to People

Large numbers of zooplankton can be killed by increases in water temperature that are common near factory outfall pipes.

FIELD MARKINGS:
Various colors and shapes, mostly translucent. Size: range in size from microscopic to jelly-fish that grow upward of 8 feet.

HABITAT:
Throughout the water column in both fresh and marine environments.

SEASONAL APPEARANCE:
Year-round; numbers tend to increase in late spring and early fall.

SENSITIVITY LEVEL:

Ocean

 Estuary

 Shoreline

Where Do I Find It?
Use these icons as a quick reference to where you might find a particular Bay species. Icons denote a species' predominant habitat; other preferred habitats are mentioned within the text.

ANATOMY OF A FISH:

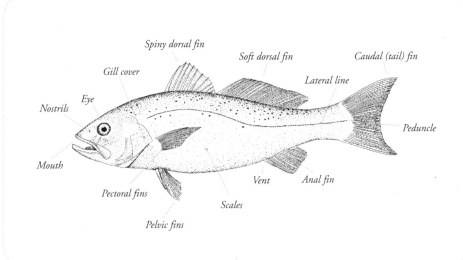

Fish

Narragansett Bay is home to more than 100 species of fish from oceanic sharks and tuna to river herring and brook trout. Some are resident species, living in the Bay year-round, while others are migratory, visiting only seasonally or accidentally. The Bay is famous for striped bass and bluefish, which are abundant from spring through fall.

Most fish belong to either the class Chondrichthyes (cartilaginous fish, also called elasmobranchs) or the class Osteichthyes (bony fish). Fish make up half of the more than 40,000 species of vertebrates in the world: 90% of those are bony fish and 10% are elasmobranchs.

Fish are cold-blooded vertebrates that breathe with blood-rich structures called gills, which extract dissolved oxygen from the water. Most fish have scales and a slimy, mucus coating that protects them from disease. Many species of fish have a swim bladder, which is a gas-filled sac that inflates or deflates to maintain neutral buoyancy in the water. Gas is taken in and expelled through the mouth and gills. Sharks, skates, clingfish and flatfish do not have swim bladders.

Most bony fish lay eggs and mate through external fertilization. The female produces thousands, sometimes millions, of eggs, which are dispersed into the water and fertilized. Most fish eggs float, but there are exceptions such as those of the winter flounder. Eggs and larvae are planktonic (free floating) until large enough to sink. Most elasmobranchs reproduce through internal fertilization and lay few eggs or egg cases, with some species birthing large, live individuals.

This section of the *Uncommon Guide* categorizes Narragansett Bay fish into groups determined by body shape and/or habitat preference. Sharks and skates are cartilaginous. The bottom and rock dwellers include species that may resemble rocks, such as the oyster toadfish, or fish like the cunner that can camouflage with the mud. Flatfish have adapted to live on the bottom with both eyes on one side of the body. Minnows are found in the shallow tidal ponds around the Bay, have small bodies and are often used as bait. Fish that are found in the open water, such as the torpedo-shaped tuna, can undergo extreme migrations at very fast speeds. Anadromous fish live in salt water but migrate to spawn in fresh water. Finally, there are those with unique shapes, such as the seahorse and various tropical fish that are accidentally transported into the Bay.

Blue Shark

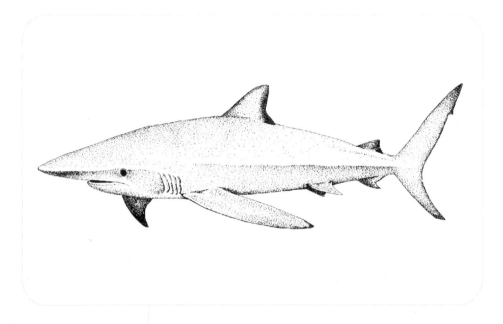

Distinguishing Features and Behaviors

Sharks, skates and rays belong to a particular group of fish called elasmobranchs, meaning their entire skeleton is made of cartilage rather than bone.

The blue shark is distinguishable from other sharks because it has a slender body and long, narrow pectoral fins. Its body is thickest at the midsection and tapers toward the head and tail. This shark has two dorsal fins – the second is about half the size of the first – and an asymmetrical tail.

This animal gets its name from the distinctive blue coloring that fades to white on its underside. When the shark dies, the blue coloring turns gray. The tough skin is composed of hard scales that are abrasive like

COLLOQUIAL NICKNAMES:
Blue Dog

SCIENTIFIC NAME:
Prionace glauca

sandpaper when rubbed against the grain. Five or more gill openings can be seen on the side of the head.

The snout of the blue shark is long with a well-rounded tip. It has large, pointed and serrated teeth that are well developed and packed close together in its mouth. The blue shark feeds on smaller fish such as herring, spiny dogfish and mackerel. It is also a scavenger, feeding on the carcasses of fish, whales and other sharks.

This shark is often seen swimming lazily at the surface, basking in the sun, and is not believed to swim to great depths. However, it is capable of attaining great speeds. The blue shark is highly migratory and found in all of the world's oceans.

Relationship to People

The blue shark is the most abundant of oceanic sharks. Its presence in Narragansett Bay waters is rare, but it has been spotted near the mouth of the Bay while migrating up and down the Atlantic coastline. This species is considered dangerous to people, but encounters are extremely unusual, and no attacks have ever been documented in Rhode Island waters.

Like most pelagic shark species, the blue shark population is severely threatened by overfishing. Recent research on the potential use of cartilage to heal cancer has lead to an increase in the harvesting of these slow-growing fish.

The practice of "finning" has decimated the population of many species of oceanic sharks, including the blue shark. Finning is the cruel practice of removing the fins from the shark and releasing it alive. Shark fin soup is considered a delicacy in many Asian countries that receive shark fins from fishermen all over the world.

FIELD MARKINGS:
Bright indigo blue back, fading to light blue-gray on the sides with a white under-side. Size: averages 7 to 8 feet long; can grow up to 12 feet.

HABITAT:
Open ocean, occasionally near mouth of the Bay.

SEASONAL APPEARANCE:
Summer.

SENSITIVITY LEVEL:

Little Skate

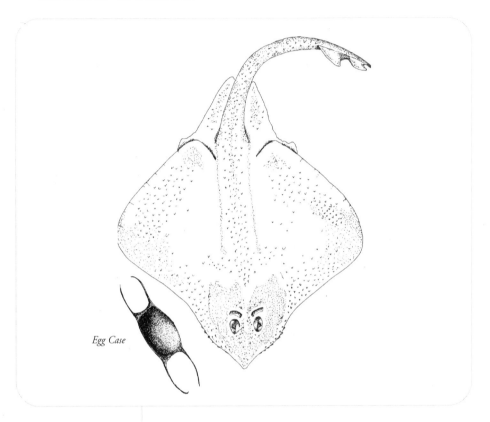

Egg Case

COLLOQUIAL NICKNAMES:
Common Skate,
Skate, Raja

SCIENTIFIC NAME:
Raja erinacea

Distinguishing Features and Behaviors

Like rays and sharks, skates belong to a group called elasmobranchs, which includes all fish with a skeleton made entirely of cartilage. One of the more common species of skate found in Narragansett Bay is the little skate. Its body is shaped like a flattened, rounded triangle and is well-adapted for life on the bottom of the Bay. The skate is armored along its back and tail with sharp spines that are used defensively, but it is not a sting ray. Females have more spines than males.

Unlike bony fish, skates lack the mechanism to pump oxygenated water over their gills. Because skates spend most of their lives on the bottom, they breathe through specialized organs called spiracles, which are slitlike openings near their eyes. Water is taken in through the spiracles, passes over the gills and then leaves the body through five pairs of gill slits underneath.

The skate has many rows of blunt teeth, resembling sand-paper, that help grind food between two well-developed jaw plates. Skates feed on a diverse diet of shellfish, crabs, sea squirts, worms, amphipods, squid and small fish.

Male skates are distinguished by two long claspers along their pelvic fins, which they use to hold onto the female and transmit sperm. Little skates copulate many times in a year. The female lays two large eggs that develop inside capsules, or egg cases, which are often found attached to seaweed. The empty black capsules wash ashore after the young skates have hatched. Resembling square coin purses with prongs at each corner, the capsules are commonly called "mermaids' purses."

Relationship to People

The little skate is quite common in Narragansett Bay. It has been increasing in numbers since 1970 and is now one of the dominant creatures in bottom-fish communities. They are frequently strung together and used to bait lobster pots. Because they reproduce slowly, their populations could decline if overexploited.

Historically, skate have not been a commercially important species, but it is becoming more popular as a food fish, frequently marketed by its Latin name, *Raja*. Skate wings are sometimes cut into small, round pieces and falsely marketed as scallops.

FIELD MARKINGS:
Light brown to gray on the back, paler toward the edges of the pectoral fins; white or gray belly. Size: averages 16 to 20 inches long, 8 to 16 inches wide.

HABITAT:
Shallow water, sandy and muddy bottoms.

SEASONAL APPEARANCE:
Spring, summer, fall.

SENSITIVITY LEVEL:

Mako Shark

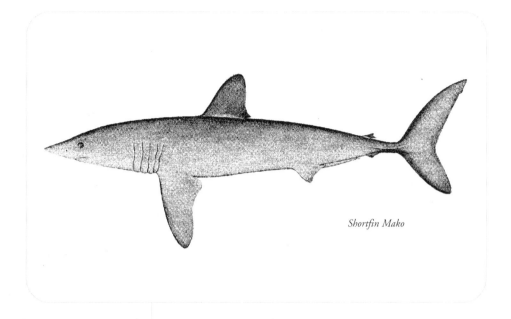

Shortfin Mako

COLLOQUIAL NICKNAMES:
Shortfin Mako,
Atlantic Mako

SCIENTIFIC NAME:
Isurus oxyrinchus

Distinguishing Features and Behavior

Sharks, skates and rays belong to a particular group of fish called elasmobranchs, meaning their entire skeleton is made of cartilage rather than bone.

The shortfin mako is an apex predator of the ocean and is noted for its agility and speed (over 30 miles per hour). When hooked, makos have been known to make spectacular, cartwheeling leaps. They have compact torpedolike bodies with a set of serrated triangular teeth in rows.

Makos are primarily piscivores, feeding on other fish such as bluefish, herring, mackerel and small tunas. They have also been known to eat some invertebrates, like squid.

Makos range the world's tropical and temperate oceans, but appear off the Rhode Island coast in summer to feed on the abundant foraging species' schools.

Relationship to People

While mako sharks have been known to attack humans, such attacks are extremely rare and none have been documented in Rhode Island.

Makos are considered excellent table fare and are targeted by recreational and commercial anglers. The state record fish weighed 718 pounds.

While the official status remains unknown, populations of makos and other large coastal sharks have declined significantly since the 1970s and are now the target of federal regulations. Mako populations have suffered worldwide as a result of fishing pressure and bycatch in the tuna and swordfish fisheries.

FIELD MARKINGS:
Gray to cobalt blue.
Size: up to 12 feet.

HABITAT:
Open ocean along the Gulf Stream, generally in deep water, but may range in close to the mouth of the Bay to chase bluefish and other bait species.

SEASONAL APPEARANCE:
Summer.

SENSITIVITY LEVEL:

Spiny Dogfish

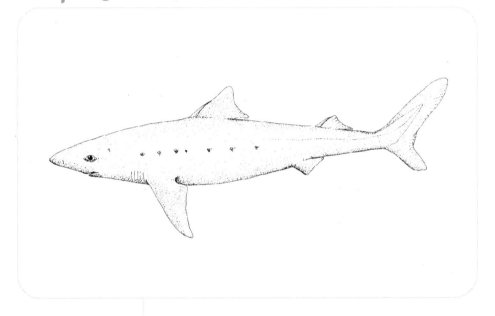

COLLOQUIAL NICKNAMES:
Dogfish, Dogshark

SCIENTIFIC NAME:
Squalus acanthius

Distinguishing Features and Behaviors

The spiny dogfish is a species of small shark with a slender, flattened head; a blunt, tapered snout and a small crescent-shaped mouth. The teeth of the dogfish are small with sharp points bending outward. The teeth are organized into several rows and are used for grinding rather than tearing. Spiny dogfish are most common in the Bay, but the smooth dogfish (Mustelis canis) can also be found.

The first dorsal fin of the spiny dogfish is somewhat larger than the second dorsal fin. Two large, sharp, mildly poisonous dorsal spines are located in front of each dorsal fin. A dogfish uses its spines defensively by curling up its body and striking at an enemy.

Dogfish skin is rough and covered by a toothlike, scale surface called dermal denticles. The skin feels smooth when rubbed with the grain of the denticles, but feels rough when rubbed against the grain.

While migrating to find food, the spiny dogfish swims in schools or packs of similar-sized individuals. The fish is extremely voracious, often scattering and destroying schools of mackerel and other fish. In addition to eating worms, shrimp, crabs and comb jellies, the spiny dogfish is one of the major predators of lobsters and large crabs.

By far, spiny dogfish are the most abundant and most commonly seen shark species in the Bay. Rather than laying eggs, this fish bears live young and can have up to six pups per litter.

Relationship to People

The spiny dogfish has been increasing in numbers in the North Atlantic, replacing overexploited groundfish stocks such as cod, haddock and yellowtail flounder. Large numbers of spiny dogfish become entangled in trawls and seine nets, doing tremendous damage to fishing gear. They are also notorious for stealing bait.

Once classified as an underutilized species, commercial fishing of spiny dogfish is at an all-time high. Because they reproduce slowly — more like mammals than fish — wild populations are susceptible to overfishing. In the past, spiny dogfish have been used for fertilizer, liver oil and as a food source, marketed as shark or scallops.

FIELD MARKINGS:
Gray or brown in color, fading to a white belly, with several white spots on its sides. Size: 2 to 3 feet long.

HABITAT:
Deeper waters near the mouth of the Bay; bottom dweller.

SEASONAL APPEARANCE:
May to November.

SENSITIVITY LEVEL:

Atlantic Sea Raven

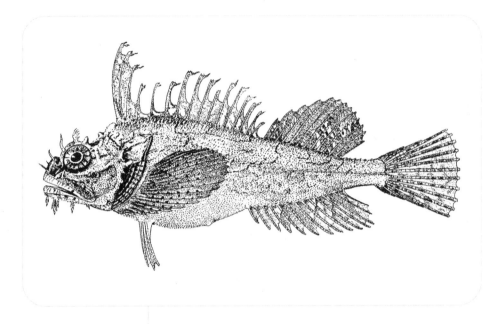

Distinguishing Features and Behaviors

The Atlantic sea raven is easily distinguishable by its head and body covering of small spines (also called "prickles"). The spines are derived from modified, platelike scales that, along with fleshy protrusions from the fish's head, help the fish to blend in with the Bay's rocky bottom. The general shape of the sea raven's spindle-shaped body and rounded pectoral and caudal fins also aids in camouflage.

There are two dorsal fins, the first with several upright spines and a ragged appearance. Pectoral fins have one spine and three soft rays. The fins and body are often maroon or brown, but color can vary according to the individual's habitat. Fins are often light and dark banded.

During the breeding season, males may display bright colors of red, orange or yellow to attract mates.

Female sea ravens lay eggs between October and December and may lay multiple sets in a season. The eggs are deposited near the base of a sponge. Larval sea ravens are predatory, feeding on the larvae of other fish. Adult sea ravens are primarily marine and are common in waters between six and 350 feet deep, with a preference for water at or below 55°F. On average, adults are between 18 and 20 inches and weigh about five pounds.

Usually feeding at night, sea ravens will eat a variety of mollusks and crustaceans, small fish and most any kind of bottom-dwelling invertebrate that is available. The primary predator of sea ravens is another bottom-dweller, the ominous-looking monkfish *(Lophius americanus)*. Upon capture, a sea raven immediately inflates its body with water as a defense to make it appear larger. If and when it is released back into the water, the sea raven will float until it is able to discharge the water from its body.

Relationship to Humans

Sea ravens are not customarily eaten by people and there is no commercial harvest of this species. Individuals are sometimes used as bait in lobster pots. Relatives of the sea raven in the Pacific Ocean are known to have poisonous spines, but the Atlantic species is harmless to people. When handled by humans, sea ravens make a humming or vibrating noise, probably as a sign of fear.

The Atlantic sea raven is very common in its range from the Chesapeake Bay north to the coast of Labrador. The population is not thought to have undergone changes in recent years.

FIELD MARKINGS:
Light to dark brown, gray and mottled to camouflage. Size: up to 25 inches and 7 pounds.

HABITAT:
Rocky or hard-bottomed areas.

SEASONAL APPEARANCE:
Year-round.

SENSITIVITY LEVEL:

Black Sea Bass

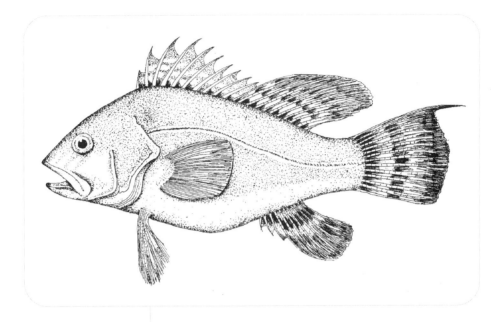

Distinguishing Features and Behaviors

Like the striped bass, the black sea bass is a "true bass" distinguishable because of its darker color. The soft spiny parts of the dorsal fin are continuous rather than separated in two parts and are marked with white spots and bands. Its caudal and pectoral fins have round edges.

Stout-bodied, the black sea bass has a moderately pointed snout and a large mouth. Its head is flat and smooth, with large eyes set high on the head. The pale centers of the scales form light, narrow stripes along the side of the fish. The male fish tends to develop a fatty hump on its back in front of the dorsal fin and is usually more darkly colored than the female.

The black sea bass is the only cold-water member of
the sea bass family, a family that includes groupers. It
can be distinguished from temperate basses by the
three spines on the gill cover and by the continuous
dorsal fin.

The majority of black sea bass undergo a sex reversal
from female to male between the ages of two and five.
Finding females over eight years old is rare, but males
can live up to 15 years.

The black sea bass is an omnivorous bottom feeder
and usually eats mollusks, crustaceans, small fish and
bottom plants. Unlike the striped bass, black sea bass
are confined strictly to salt water.

Although black sea bass are solitary and often territorial,
they gather in large groups in deep water to spawn in
late spring.

Relationship to People

The black sea bass is an important bottom-fish species
of the mid-Atlantic, taken for both food and sport. The
number taken recreationally in New England waters is
generally greater than the number taken commercially.

Although the number of black sea bass landings has
not decreased dramatically in the last 15 years, there
is rising concern about the health of the stocks, as the
average size of individual fish has become smaller.

FIELD MARKINGS:
Dark smoky gray
to dusky brown or
blue-black, with pale
sides. Size: 1 to 2
feet long; up to 5
pounds.

HABITAT:
Rocky bottoms;
near reefs, wrecks
and oyster bars.

SEASONAL APPEARANCE:
May to November.

SENSITIVITY LEVEL:

Cunner

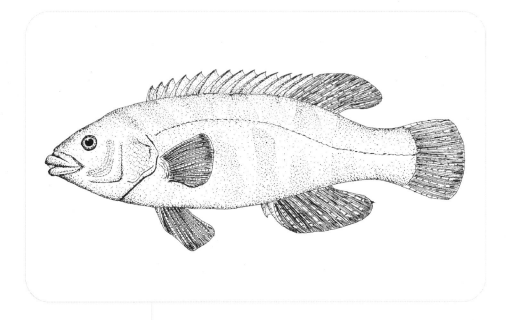

COLLOQUIAL NICKNAMES:
Choggie

SCIENTIFIC NAME:
Tautogolabrus adspersus

Distinguishing Features and Behaviors

The cunner is a small, slender fish that belongs to the wrasse family of fish. It is characterized by a single, long dorsal fin, with sharp spines forward and soft rays in the rear. Closely related to and often incorrectly identified as a tautog, the cunner is generally smaller than a tautog, less stout, and has thinner lips. It also has distinct iridescent blue streaks running from its mouth back to its gill cover.

The cunner has large scales and tough skin with a vertically flattened body. Its flat-topped head has a pointed snout and a small mouth, generally exposing several sharp teeth. The cunner's tail fin is blunt with rounded corners.

Cunner live near the coastline and are usually found inhabiting eelgrass beds and swimming near piers, docks and among rocks. Although it rarely travels into brackish water, it is occasionally seen in tidal creeks. Some cunner live together in small groups, but they do not school.

Although the cunner lives in Narragansett Bay year-round, it lies dormant in the mud — among rocks or eelgrass beds — during the winter season.

Cunner are aggressive omnivores as well as scavengers. They feed on barnacles, mollusks, shrimp, crabs, amphipods, small fish and almost any other available food, including eelgrass.

It is difficult to determine the age of a cunner simply by looking at its size, as growth rates differ among individuals, and females often grow larger than males.

Relationship to People

At one time, cunner was considered a favorite food fish, but it is no longer commonly sought after commercially. It is not a popular game fish but is occasionally caught recreationally. Cunner can be a nuisance to fishermen because they often steal bait.

If you look carefully, it is possible to view cunner in the larger, near-shore tide pools along Narragansett Bay. They hide under rocks and around large clumps of Irish moss and other seaweeds, well-camouflaged due to their color.

FIELD MARKINGS:
Green-gray with some blotching; can change color to blend in with the bottom. Electric blue streaks run from mouth back to gill cover. Size: up to 10 inches long.

HABITAT:
Along the coastline, just below the tide mark among eelgrass, pilings, Irish moss and rocky shores.

SEASONAL APPEARANCE:
Year-round.

SENSITIVITY LEVEL:

Naked Goby

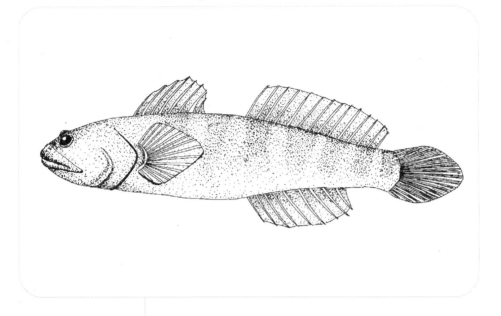

COLLOQUIAL NICKNAMES:
Goby

SCIENTIFIC NAME:
Gobiosoma bosci

Distinguishing Features and Behaviors

Gobies are bottom-dwelling fish resembling small lizards. They have large eyes set close together on top of their heads. Gobies are distinguishable by a round tail, two separate dorsal fins and fused pelvic fins. These pelvic fins act as suction cups and are used to cling to rocks and shells. The naked goby is a small fish that inhabits areas in and around shell and rock communities. Having no scales, they are aptly named "naked" goby and are smooth to the touch.

The goby family has more species than any other family of fish, with more than 2,000 varieties accounted for in the world's oceans. Most of these species live in tropical waters, but several are found in the waters of the North Atlantic. Of all the temperate species of the goby, the naked goby is the most abundant in Narragansett Bay.

Generally a solitary and reclusive animal, gobies inhabit shallow marshes, mud flats and oyster reefs, often hiding inside empty clam and oyster shells. Female gobies lay their eggs inside dead oyster shells and leave the male to guard the nest until the eggs hatch.

The naked goby feeds on worms and amphipods and is preyed upon by eels, shrimp and larger fish. In the winter, the goby becomes sluggish, ceasing to feed or swim, and seeks shelter from predators.

Relationship to People

Not commercially valuable due to its small size and solitary behavior, the goby generally escapes human interest. However, they often use human discards such as cans, bottles and tires as artificial safe havens from predators.

Many species of goby are popular as aquarium fish. Some are brightly colored, while others are transparent. Certain species of gobies are known as "cleaners" and actively remove parasites from larger fish.

FIELD MARKINGS: Dark greenish-brown on top and pale below, with eight to nine dark vertical bars along its side. Size: up to 2.5 inches long.

HABITAT: Protected coastal waters, underwater vegetation and seagrass beds, tidal fresh waters.

SEASONAL APPEARANCE: Year-round.

SENSITIVITY LEVEL:

Northern Sea Robin

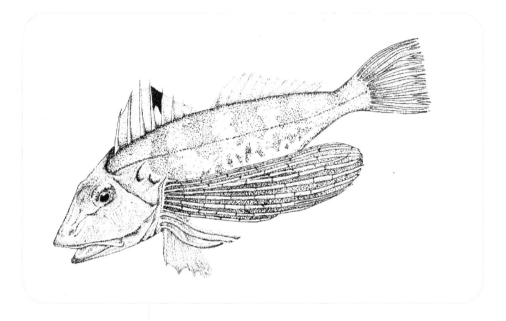

COLLOQUIAL NICKNAMES:
Common Sea Robin

SCIENTIFIC NAME:
Prionotus carolinus

Distinguishing Features and Behaviors

The northern sea robin is distinguishable by a large, spiny head and tapering body. It is easily identified by its rounded, fanlike pectoral fins, which are so large that, when laid back, they resemble wings.

The three lower rays of the sea robin's pectoral fins are long, broad feelers used for walking along the bottom. Sea robins use these lower fin rays as sense organs to stir up bottom sediments and find food.

The head of the northern sea robin is encased in bony plates that act as a shovel to dig up invertebrates from the mud. The front part of its upper jaw is concave, and there is a small spine in its nostrils.

The northern sea robin's eyes are a distinctive peacock blue. The features of its head distinguish a sea robin from a similar-looking fish, the sculpin. Another common species, the striped sea robin *(Prionotus evolans)*, is usually larger than the northern sea robin, growing up to 18 inches. The striped sea robin has a wider mouth, flatter head and multiple stripes above the lateral line.

The sea robin feeds on a wide variety of invertebrates, including shrimp, crabs, amphipods, squid, bivalve mollusks and segmented worms. It has also been known to bite readily on any bait, suggesting a non-selective feeding habit.

Sea robins typically inhabit areas of hard, smooth ground. These active swimmers are sometimes found close to the surface, but when threatened, they bury themselves in the sand, revealing only their eyes and the top of their heads.

Relationship to People

Sea robins are plentiful in southern New England. Since they are well-known bait stealers, they can be a nuisance to fishermen. Some people find these bottom-dwellers good to eat and harvest them mainly by hook and line. Their numbers in Narragansett Bay have been slowly declining over the last several years, which may be due in part to incidental capture by increased commercial fishing.

These fish are interesting to encounter, often producing an audible "croak" as a defense mechanism when held out of the water; they do this by vibrating muscles against an air-filled sack.

FIELD MARKINGS:
Combination of red, gray and brown, with dark blotches along its back; underside is dirty white or pale yellow. Size: 12 to 16 inches long.

HABITAT:
Smooth, hard-packed bottom of open Bay and among rocks.

SEASONAL APPEARANCE:
May to October.

SENSITIVITY LEVEL:

Oyster Toadfish

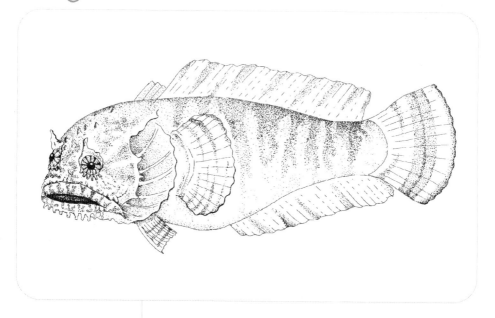

COLLOQUIAL NICKNAMES:
Toadfish, Dowdy

SCIENTIFIC NAME:
Opsanus tau

Distinguishing Features and Behaviors

The oyster toadfish is an unusually shaped, large-headed fish that typically lives along oyster reefs and vegetated muddy bottoms. It has scaleless skin and is covered instead with a thick mucus and, possibly, warts, making it easy to understand why it earns the name "oyster toadfish." Out of water, this fish feels soft and squishy.

Toadfish have a tapered body with a plump belly and a large, flat head that tapers to a thin tail. Its nose is rounded, and it has a tremendous mouth with large, blunt teeth. Its lips and eyes are surrounded by thick, fleshy flaps of skin. There are two sharp spines, located on the gill covers, which the toadfish uses for defense. The ventral fins are located underneath its throat in front of the gill openings; the fins stretch out like fans.

A voracious omnivore, the oyster toadfish feeds aggressively on oysters, worms, shrimp, amphipods, crabs, mollusks, squid and small fish. The strong teeth and jaws of the toadfish are capable of crushing the hard shells of mollusks and are often used to fight with other oyster toadfish. They snap viciously when caught.

Oyster toadfish are quite vocal. To attract a female during spawning season, the male emits a loud foghorn-like call that can be heard underwater for great distances. When handled out of water, toadfish grunt. The female lays her eggs in crevices, under submerged wood and, sometimes, even in discarded tin cans. After fertilization, the female leaves the male to guard the nest. For about a month, the male cleans the nest and uses his fins to fan the eggs until they hatch. The male continues guarding the young for three to four weeks after the eggs have hatched, keeping watch over the juveniles even after they become free swimmers.

Relationship to People

Oyster toadfish appear to be well-adapted to living among pollution and litter and have been found inside tires or cans submerged in the Bay.

Because of its size, sensitivity to pollution and ability to live out of the water for a long time, the oyster toadfish serves as an important animal for marine research.

FIELD MARKINGS:
Blotchy, olive brown body, fading to pale below. Capable of changing color to match the bottom.
Size: up to 12 inches long.

HABITAT:
Sandy, rocky and muddy bottoms on oyster reefs, shallow water, among eelgrass, hollows or dens.

SEASONAL APPEARANCE:
Year-round.

SENSITIVITY LEVEL:

Scup

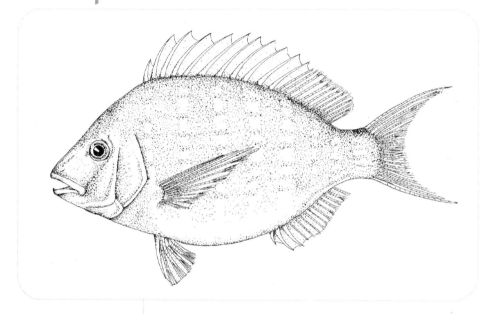

Distinguishing Features and Behaviors

The scup is a medium-sized fish with a deep, vertical compressed body. Its scales are iridescent, often reflecting the color of the rainbow. Scup have small mouths with strong jaws and pointed teeth used to crush small mollusk shells. Its dorsal fin is composed of sharp spines that make up more than half the entire fin length.

Scup are migratory species that travel in schools of similarly-sized fish. The thin, dark lateral line acts as a sense organ, helping the fish to detect tiny movements in the water. This allows them to sense danger, find food and move quickly in a school without bumping into each other. Scup are bottom feeders, often gathering near rocks and submerged pilings to feed on barnacles, mollusks, worms and other invertebrates.

Sensitive to cold water temperatures, scup will move offshore into deeper waters during the winter. If caught in shallow waters during extremely cold weather, these fish often die. During the summer months, they tend to gather near the coastline and are never found more than a few miles offshore.

Relationship to People

Scup are a favorite sport fish for anglers in the Bay. Second only to bluefish in number of pounds landed by recreational fishermen, they are fished with worms, clams or squid from June to August over sandy bottoms. Scup are also an important commercial species, harvested in great numbers each year. This species is commonly exported to Japan.

At one time, scup and winter flounder comprised the majority of fish species in Narragansett Bay. As a result of overfishing and habitat destruction, scup are now considered to be an overexploited species.

FIELD MARKINGS:
Silvery-gray with faint, irregular, dark bars and pale blue flecks on its sides. Size: 4 to 10 inches long.

HABITAT:
Sandy and rocky bottoms, open waters.

SEASONAL APPEARANCE:
May to October.

SENSITIVITY LEVEL:

Spotted Hake

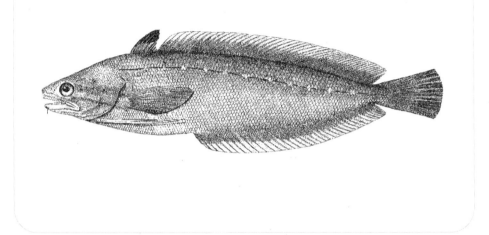

SCIENTIFIC NAME:
Urophycis regius

Distinguishing Features and Behaviors

Spotted hake are a bottom-dwelling fish, camou-flaged by a dark, dull brown color on their dorsal side. One of the smaller species of hake, they are built like predators. They have streamlined bodies, large well-developed eyes and strong teeth, and they catch their food by ambush. Like other members of the cod family, hake have a distinctive fleshy barbel on their chin. Their diet includes shrimp, squid, crabs, bivalves and fish, including members of their own species.

Spotted hake have earned their name because their lateral line, which is a darker brown than their body, is interrupted by white spots. Additional features that distinguish them from other hake species are: the outer half of their first dorsal fin is black and their pectoral fins reach back as far as the beginning of the anal fin.

Although non-migratory, spotted hake may move into deeper waters to overwinter. Juveniles spend part of their time in estuarine habitats like Narragansett Bay, while adults can be found further offshore on the continental shelf and slope waters in cold-temperate and arctic waters. The most frequent method of capture and observation has been by bottom trawl.

Relationship to People

Hake species are part of the cod family (gadidae, which includes cod, haddock and pollock) and are an ecologically and commercially important species to fisheries in the northwest Atlantic Ocean. Although spotted hake are relatively abundant, they are not as commercially important as red, silver or white hake, due to their small size and lesser quality meat. They are aggressive, and thus, are often accidentally caught by recreational fishermen targeting other species.

FIELD MARKINGS:
Dull brown; darker above than below, with a lateral line darker than its body color and interrupted by white spots. The outer half of its first dorsal fin is black. Size: up to 16 inches long.

HABITAT:
Under rocks and benthic vegetation along the continental shelf.

SEASONAL APPEARANCE:
Year-round.

SENSITIVITY LEVEL:

Tautog

COLLOQUIAL NICKNAMES:
Blackfish, Chinner

SCIENTIFIC NAME:
Tautoga onitis

Distinguishing Features and Behaviors

Tautog are heavy, stout fish with a broad tail and a high, arched head. They are the northern relative of the family of wrasses that are common in tropical waters. Tautog are related to and often confused with another species of wrasse — the cunner. The tautog has a blunt snout with a small mouth, thick lips and strong conical teeth. They have a scaleless cheek region that is smooth to the touch. Their dorsal fin extends the length of the back and has sharp spines.

Tautog become blacker as they grow older, and their coloring also varies depending on the local bottom habitat. The distinguishing feature of the adult male tautog is the large protruding forehead. Mature males are often referred to as "chinners" because of the white patch on the chin.

Tautog feed entirely on invertebrates, including crabs, mussels, mollusks, shrimp, amphipods and sea worms, using their strong back teeth to crush any hard shells. These fish are not active swimmers and, when not feeding, they often gather in groups under the safety of a ledge or hole in the rocks, sometimes lying on their sides. Although tautog are active during the day, they remain close to cover. At night they are quiet and inactive, hiding from predators.

Juvenile tautog stay near the sites where they were hatched and are frequently found in eelgrass beds where invertebrates are abundant. The adults gather around rocky bottoms, ledges, pilings and submerged wrecks.

Relationship to People

Although they are sometimes sold commercially, most tautog are caught recreationally. Taken by rod and reel and by spear from May through October, tautog are an important sport fish in Narragansett Bay. There is a "live fish" fishery in Rhode Island where living adult tautog are captured for restaurants, and customers can select live fresh fish from the tanks.

Increased pressure by commercial gill-net fisheries and recreational fishermen has resulted in serious decline of stocks. In addition, tautog grow slowly, taking a long time to reach sexual maturity. This makes it difficult for the stocks to rebound quickly when overfished. In many states, including Rhode Island, a minimum size limit for recreational fishing has been imposed to help maintain the population.

FIELD MARKINGS:
Males and older fish are uniformly olive green, dark chocolate or black in color with irregular mottling along the side. Females and young tautog are paler with large mousy brown and gray mottling on the sides. Size: up to 22 inches long.

HABITAT:
Open water near rocky shores, pier docks, breakwaters, mussel beds.

SEASONAL APPEARANCE:
Year-round, most common from April through November.

SENSITIVITY LEVEL:

Atlantic Silverside

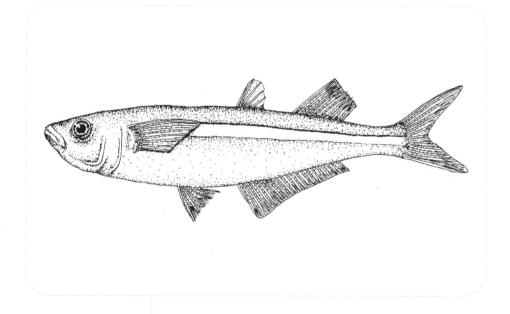

COLLOQUIAL NICKNAMES:
Silverside, Baitfish

SCIENTIFIC NAME:
Menidia menidia

Distinguishing Features and Behaviors

The Atlantic silverside is a long, slender and thin-bodied fish with two dorsal fins, a round white belly, and large scales. It has a short head with large eyes and a small mouth. Along each side, from the pectoral fin to its caudal fin, is a distinct silver band outlined by a narrow black stripe. Silversides resemble anchovies, differing mainly in mouth size.

Atlantic silversides congregate in large schools that usually consist of similar-sized fish. They are found along the shore, often within a few feet of the water's edge. The silverside is commonly seen swimming among submerged grasses in brackish waters where streams and rivers meet the sea.

In summer, they are rarely found in water deeper than a few feet, but will descend to greater depths in the winter to avoid the cold temperatures of shallow water.

This omnivorous fish feeds on zooplankton, copepods, shrimp, amphipods, young squid, worms, insects and algae. They are favored prey for larger predatory fish, such as mackerel, striped bass and bluefish, and are eaten by birds such as terns and cormorants.

The Atlantic silverside frequently interacts with another similar species, the inland, or waxen, silverside *(Menidia beryllina)*. The inland silverside is smaller and stouter than the Atlantic silverside and duller in color. These two species occasionally school together and can coexist without significant competition for habitat, food or space.

Relationship to People

The Atlantic silverside is the most abundant fish in Narragansett Bay. Silversides are an important food source for young bluefish, mackerel, striped bass and most shorebirds. Silversides are used to bait eel pots. Flocks of birds hovering and diving usually indicate the presence of silversides being fed upon by a school of juvenile bluefish or stripers.

The silverside is a common subject for scientific research because of its sensitivity to extreme environmental conditions such as low oxygen levels, drastic tempera-ture changes and contaminants in the water.

FIELD MARKINGS:
Translucent gray-green above and pale below with thick, dark brown speckles on its upper sides. Distinct silver band along the sides. The top of the head, nose and chin are dusky gray. Size: up to 5.5 inches long.

HABITAT:
Sandy or gravel shores, brackish estuaries, salt water river mouths.

SEASONAL APPEARANCE:
Year-round.

SENSITIVITY LEVEL:

Fourspine Stickleback

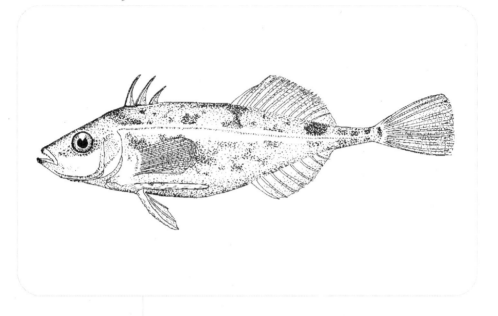

COLLOQUIAL NICKNAMES:
Stickleback

SCIENTIFIC NAME:
Apeltes quadracus

Distinguishing Features and Behaviors

The stickleback is a tiny fish readily identified by the presence of two or more short spines along its back, just in front of its dorsal fin. These spines can be raised or retracted at will, and are usually used for defense. The number of dorsal spines found on the back determines the species of stickleback.

The fourspine stickleback is the most common in the Bay, but two other species, the threespine *(Gasterosteus aculeatus)* and ninespine stickleback *(Pungitius pungitius)* are also found. Distinguishing between the threespine and fourspine is sometimes difficult because the last spine is actually a part of the soft dorsal fin. The threespine stickleback only has two visible, single spines, and the fourspine stickleback has three that are visible.

The fourspine stickleback is scaleless, while the threespine stickleback has armor plates along the sides of its body. Both species are small and slender with a flattened belly. The ninespine has anywhere from seven to twelve spines.

The diet of sticklebacks is composed of copepods and other small planktonic crustaceans.

The stickleback spawns in early May through late July. The male carries weeds, leaves and grasses that are cemented together in its mouth with mucus threads. The fish uses these materials to build a nest, then leads its mate there to spawn. The nest of the threespine stickleback is round, while the fourspine's nest is cone-shaped.

After the female lays eggs in the nest, the male fertilizes them and guards the nest during the six-day incubation period. The male stickleback removes dead eggs from the nest and returns any live eggs that fall out. Males protect the newly hatched fish until they can survive on their own. Most males die after the spawning season, and those that do survive return to sea with the females.

Relationship to People

A common fish in salt marshes, sticklebacks are often found in the same areas where mummichogs and silversides are present. Although it can often be found in fresh water, the fourspine stickleback is primarily a brackish or saltwater fish.

Sticklebacks have no commercial value in the United States, but in some European and Scandinavian countries they are so plentiful that they are harvested for the oil present in their bodies.

FIELD MARKINGS:
Olive or green-brown with dark mottling that alternates below the lateral line. Silver-white belly with red ventral fin membranes. Males are usually darker than females. Size: 1.5 to 2.5 inches long.

HABITAT:
Salt marshes, tidal creeks, underwater vegetation, near-shore environments.

SEASONAL APPEARANCE:
Year-round.

SENSITIVITY LEVEL:

Killifish

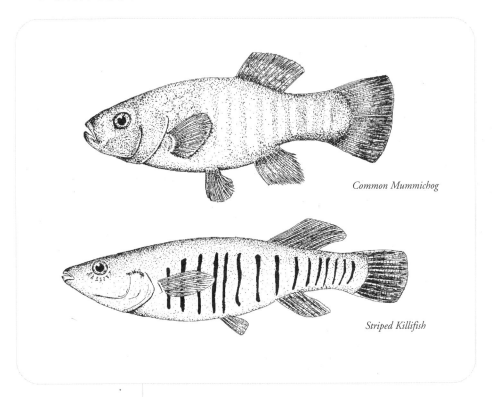

Common Mummichog

Striped Killifish

COLLOQUIAL NICKNAMES:
Mummichog, Mummy,
Chub, Minnow

SCIENTIFIC NAME:
Common Mummichog:
Fundulus heteroclitus
Striped Killifish:
Fundulus majalis

Distinguishing Features and Behaviors

The name "killifish" is often used to describe small fish of the *Fundulus* genus, commonly known as mummichogs. In Narragansett Bay, at least two species are present: the common mummichog and the striped killifish.

The body shape of both species is similar. They have a round back and belly, a soft dorsal fin situated far back on the body, a thick, rounded tail, a head that is flat on top between the eyes, a blunt snout and a tiny mouth. The head of the striped killifish is slightly longer and thinner than the common mummichog.

Although generally the same size, the striped killifish is slightly larger and more slender than the common mummichog. During breeding season, the male striped killifish becomes brilliantly colored: its back turns black, its sides become golden orange, and its fins turn bright yellow. Although coloring also intensifies for the common mummichog, it remains paler in comparison.

"Mummichog" is a native peoples word that literally means "going in crowd," a name that aptly describes the schooling activities of the mummichogs. Mummies rarely stray more than 100 yards from the shoreline.

Mummichogs are omnivores and scavengers, and feed in salt marshes on phytoplankton, mollusks, crustaceans and dead fish.

Relationship to People

Both the common mummichog and the striped killifish are used as bait in recreational fishing. Because they feed on mosquito larvae, they are a safe substitute for harmful pesticides used to control mosquitoes.

The mummichog is able to withstand foul water with low oxygen levels and extreme variations in salinity — conditions in which most fish species would quickly perish. They are often found stranded in tide pools or puddles of water from the receding tide. If the water dries up before the next tide comes in, mummichogs can flop head-over-tail to reach the water. They winter in tidal creeks where the salinity conditions are lower than in marshes. There, they are often found in a sluggish state, buried in up to eight inches of mud.

FIELD MARKINGS:
Common mummi-chog: olive green to blue-back, white belly; males have irregular silvery bars or mottling.

Striped killifish: irregular vertical black bars along the sides. Females have three horizontal bars, with the lowest bar divided into two. Size: both range from 5 to 7 inches long.

HABITAT:
Sheltered salt marshes, tidal creeks, brackish water.

SEASONAL APPEARANCE:
Year-round; more abundant in spring, summer, fall.

SENSITIVITY LEVEL:

Sheepshead Minnow

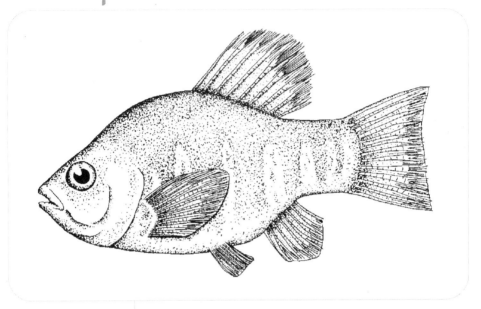

COLLOQUIAL NICKNAMES:
Variegated Minnow

SCIENTIFIC NAME:
Cyprinodon variegatus

Distinguishing Features and Behaviors

The sheepshead minnow is a small, thick-bodied fish with a high, arched back and a flat-topped head. These fish are found in shallow, brackish inlets and coastal ponds, often in the company of mummichogs and silversides. Sheepshead minnows are immediately distinguished from mummichogs by their thicker body shape, which is nearly half as tall as it is long, and by their thick, square tails. The body and head are covered with large, rounded scales, and its small mouth holds large, wedge-shaped teeth with tricuspid cutting edges.

On average, males are larger than females. The male has a black bar along the square edges of its tail, while the female has an obvious dark spot on the back of the dorsal fin. During the spawning season, males become an iridescent blue with a dark orange belly. Young sheepshead minnows are more slender than adults and have irregular bands on their sides.

The sheepshead minnow is omnivorous, feeding on plants, invertebrates and other fish. They are quite aggressive and can injure and kill fish much larger than themselves by repeatedly slashing a victim with their sharp teeth.

During the winter months, sheepshead minnows burrow in the mud, lying dormant until spring.

Relationship to People

Found in salt marshes, sheepshead minnows have a high resistance to environmental extremes. Like mummichogs, they can withstand the changing levels of oxygen, temperature and salinity that are common in coastal salt marshes. Sheepshead minnows are often used as bait in recreational fishing.

FIELD MARKINGS:
Olive green with pale yellow and white belly. Males turn bright irides-cent blue in the spring. Size: less than 1.5 inches long.

HABITAT:
Shallow waters around inlets, harbors, salt marshes.

SEASONAL APPEARANCE:
Year-round.

SENSITIVITY LEVEL:

Atlantic Bluefin Tuna

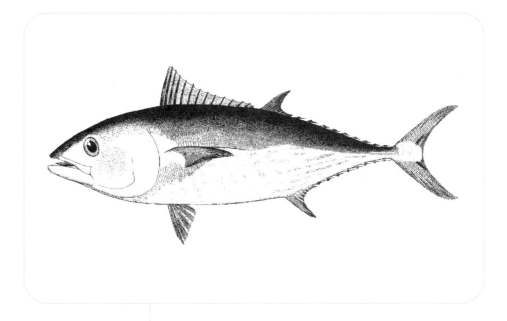

COLLOQUIAL NICKNAMES:
Bluefin, Giant Tuna, Horse Mackerel, Great Albacore

SCIENTIFIC NAME:
Thunnus thynnus

Distinguishing Features and Behavior

The Atlantic bluefin tuna is silver-bodied with a dark blue back. It sometimes has vertical stripes and iridescent colors.

Tuna are amazingly strong and fast swimmers and are among the top predators in the ocean. Bluefin can grow to be the legendary "Giant Tuna," and off the Rhode Island coast, they range from "school tuna" (ten to 100 pounds) to "mediums" (100-300 pounds) to "giants" (300-plus pounds). Rarely found in Narragansett Bay proper, bluefin are found with some regularity in Block Island Sound.

Bluefin tuna migrate great distances, although their exact pattern is still not well understood. Tuna range from the Gulf Stream current northward along the eastern coast of North America from spring through fall, chasing smaller forage fish. They frequently travel in schools, though larger fish may be solitary. Bluefin tuna may migrate between the east and west Atlantic Ocean as well.

When feeding on baitfish, schools of tuna may jump or porpoise through the water, creating massive "boils" on the surface. Their prey varies, but includes herring, mackerel and bluefish, among others.

Relationship to People

Bluefin tuna is prized as a game fish and a food fish — so much so that these magnificent creatures are becoming increasingly scarce. The fact that they migrate so widely through international waters makes their management particularly challenging.

Throughout Rhode Island's history, the bluefin tuna fishery was a staple of late summer. For decades, the Galilee Annual Tuna Tournament on Labor Day weekend would draw tourists who would gather to catch a glimpse of a "giant" at the dock. Fish would be weighed and then immediately cleaned and flown to Japan where they would draw very high prices for use as sushi.

Since the 1990s Atlantic bluefin tuna populations have declined dramatically, and their management has become an international priority.

FIELD MARKINGS:
Silver-bodied with dark blue backs; may have vertical stripes and iridescent colors. Size: up to 1200 pounds and more than 9 feet long.

HABITAT:
Deep water toward the continental shelf.

SEASONAL APPEARANCE:
July through October.

SENSITIVITY LEVEL:

Atlantic Green Bonito
and False Albacore

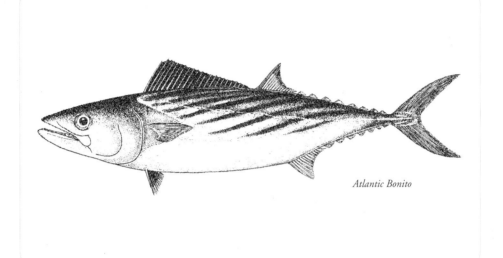

Atlantic Bonito

COLLOQUIAL NICKNAMES:
*Bonito, Greenies,
Little Tunny*

SCIENTIFIC NAME:
*Bonito: Sarda sarda
Albacore: Euthynnus
      alletteratus*
      .

Distinguishing Features and Behavior

Nicknamed "Atlantic speedsters" for their impressive strength and speed, both the bonito and false albacore are favorite game fish species. They resemble miniature tuna and are related as members of the mackerel family.

Both species travel in large schools, creating frenzied splashes as they blitz small baitfish such as juvenile menhaden and silversides.

The bonito is silver underneath and green to steely-blue on its back with distinctive dark horizontal bars on the sides. The false albacore is also silver and green with a more mackerellike, criss-cross pattern above the lateral line. Both commonly range from four to twelve pounds, with some reaching 20 pounds, but fish this large are rare.

With a single row of long sharp teeth, bonito adults can be cannibalistic, although they also eat smaller schooling fish, squid and shrimp. False albacore have smaller teeth and feed opportunistically on fish, squid, crustaceans and tunicates.

Relationship to People

Both the bonito and false albacore are edible, but oily and strong-tasting; the bonito is considered the better meal of the two.

Neither the bonito nor the false albacore support a dedicated commercial fishery in Narragansett Bay, and neither stock is considered to be overfished. Their numbers vary from year to year, but both seem to be reasonably stable.

FIELD MARKINGS:
Bonito is green to steely-blue on the back with distinctive dark horizontal bars on the sides.

False albacore is silver and green with a criss-cross pattern above the lateral line. Size: both range from 4-12 pounds.

HABITAT:
Open ocean and coastline.

SEASONAL APPEARANCE:
July through October.

SENSITIVITY LEVEL:

Atlantic Moonfish

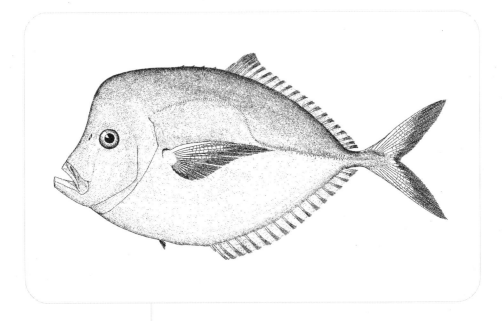

COLLOQUIAL NICKNAMES:
*Lookdown, Mother-in-law
Fish, Dollarfish, Flatjack*

SCIENTIFIC NAME:
Selene setapinnis

Distinguishing Features and Behaviors

The Atlantic moonfish is a deep-bodied, short and extremely laterally compressed (very thin) fish in the jack family. The facial profile of the moonfish is nearly vertical and concave, and their caudal fin is deeply forked. As juveniles, they have an elongated black spot on their midside.

Only juvenile moonfish are found in Narragansett Bay, and they live on muddy bottoms in brackish estuaries or other coastal marine waters. These waters are ideal because Atlantic moonfish feed on smaller fish, worms and crustaceans, such as small crabs and shrimp.

Adults spawn offshore, and their eggs and larvae are transported by the Gulf Stream current. Despite their sub-tropical nature, Atlantic moonfish have been caught all over the western North Atlantic, mostly in bottom-trawl sampling gear.

These fish, considered good swimmers, tend to school together, especially when young. As they get larger, they may be encountered in small groups or pairs.

A similar species found in the Bay is the lookdown *(Selene vomer)*, which has a steeper forehead and a sharper nose point. Like the Atlantic moonfish, lookdown are only found in the Bay as juveniles, when they have long impressive (often yellow) filaments that flow from their dorsal and anal fins.

Relationship to People

Adult Atlantic moonfish are a highly valued commercial fishery, as they are good food fish when marketed fresh. Despite the high quality of their meat, there have been reports of ciguatera poisoning in the Caribbean from eating these species. The poisoning is caused by the consumption of tropical fish species that have accumulated toxins in their muscles due to ingesting certain phytoplankton species (mostly dinoflagellates).

Atlantic moonfish are respected as game fish and are said to put up quite a fight on hook-and-line. They are also used as aquarium fish due to their iridescent coloration.

FIELD MARKINGS:
Silvery gray-blue and iridescent, often with brassy highlights. Size: up to 24 inches and 10 pounds.

HABITAT:
Near the bottom, occasionally schooling near the top. Juveniles are found on muddy bottoms in brackish estuaries.

SEASONAL APPEARANCE:
Late summer through early fall.

SENSITIVITY LEVEL:

Bluefish

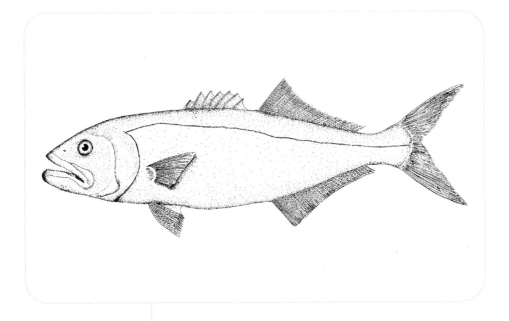

COLLOQUIAL NICKNAMES:
*Blues, Snapper Blues,
Skipjack*

SCIENTIFIC NAME:
Pomatomus saltatrix

Distinguishing Features and Behaviors

The bluefish is a pelagic fish species that migrates north and south along the Atlantic coast. Bluefish found in Narragansett Bay are part of a larger population that migrates seasonally from Maine in the summer months to Florida in the winter. Bluefish frequent the Bay in early summer, departing south by late November.

The bluefish has a long, stout torpedolike body and a forked tail. This shape makes it a fast, hydrodynamic, efficient swimmer that can travel long distances. It is a voracious predator with a large mouth and strong, sharp, triangular teeth. It feeds opportunistically on fish, squid, crabs, lobsters and shrimp. Using a slashing attack style, schools of bluefish often kill more prey than they can eat. Over 70 species of finfish, including alewives, butterfish, silversides and juvenile bluefish, have been identified in bluefish stomach contents.

During its first few years, the bluefish feeds intensely and grows rapidly. Young bluefish, called snappers or skipjacks, enter the Bay in June as small as two inches in length and grow up to ten inches by the time they leave in late September. Juveniles seek protected waters such as estuaries and salt ponds for feeding and safety from predators.

Relationship to People

Bluefish often feed on large schools of baitfish, creating a disturbance known as "blitzing" in the water. In an attempt to escape, small fish leap into the air while the bluefish churns up the water with its tail and snapping jaws. During these feeding frenzies, the water actually looks like it is boiling.

Due to the strong fight it puts up, the bluefish is a favorite catch of recreational fishers along the Atlantic coast. Many also enjoy the flavor of the oily muscle. Although 80-90% have been taken recreationally over the last ten years, the bluefish is also an important commercial fish species. Fishing for skipjacks is a popular sport in Rhode Island during summer and fall. Current regulations allow fisherman up to ten fish per day with no size limit.

In recent years, there has been a general decline in bluefish abundance. This is evidenced by fewer landings and a decline in spawning-stock populations.

FIELD MARKINGS:
Distinct greenish-blue with silver sides and a characteristic dark spot behind its pectoral fin. Size: 30 inches long; can reach lengths of 45 inches and weigh up to 25 pounds.

HABITAT:
Along the shoreline to deep water; juveniles near shore.

SEASONAL APPEARANCE:
May to late November.

SENSITIVITY LEVEL:

Butterfish

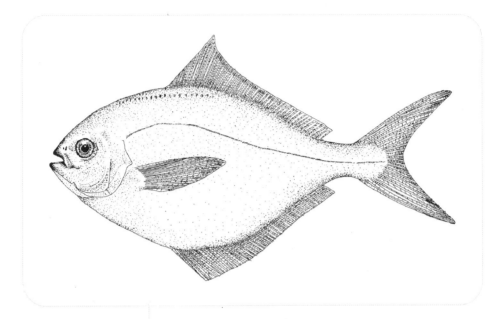

COLLOQUIAL NICKNAMES:
Shiner, Butters, Dollarfish

SCIENTIFIC NAME:
Peprilus triacanthus

Distinguishing Features and Behaviors

The butterfish is a small, round fish distinguishable by its thin, deep body and lack of pelvic or ventral fins. The butterfish has a soft-rayed dorsal fin running along the length of its back and an anal fin almost as long. Its tail is deeply forked, and the pectoral fins are long and pointed.

The butterfish has a small mouth with a single row of weak teeth and a concealed upper lip. Its snout is heavy and rounded, and the large eyes are rimmed with fatty tissue. Butterfish scales are quite small and will easily slough off when touched.

Butterfish travel in large schools, preying on small pelagic fish, shrimp, squid and sometimes jellyfish and comb jellies. Young butterfish are often seen taking shelter among the tentacles of sea nettles and other jellyfish, apparently immune to the toxins in the stinging tentacles.

The butterfish matures after the first year but rarely lives past the age of three. The young fish stays close to the shore during the first year of its life and prefers areas of high salinity to the fresher estuarine waters. Butterfish found in Narragansett Bay are part of a larger population of butterfish that migrate along the Atlantic coastline from southern New England to Cape Hatteras. They migrate out of the Bay to deeper waters in late fall as water temperatures cool. When in the Bay, the butterfish swims near the surface, particularly over sandy-bottom habitats.

Relationship to People

Butterfish are an important food fish and have been harvested commercially since the 1800s. In the early 1900s they were used primarily as fertilizer, but were then discovered to be suitable for eating as well.

Often used as bait in recreational fishing, butterfish are a favorite food source for large game fish such as tuna. They are considered underexploited as a fish resource, and are currently under a mid-Atlantic Fishery Management Plan that limits the amount of fish that can be harvested each year in an effort to prevent overexploitation of the species.

FIELD MARKINGS:
Grayish-blue on back with silvery sides and belly and numerous irregular dark spots. Size: 6 to 9 inches long; weighs less than half a pound.

HABITAT:
Sheltered bays and estuaries, sandy bottoms along the continental shelf; prefer areas of high salinity.

SEASONAL APPEARANCE:
Late April to August.

SENSITIVITY LEVEL:

Squeteague

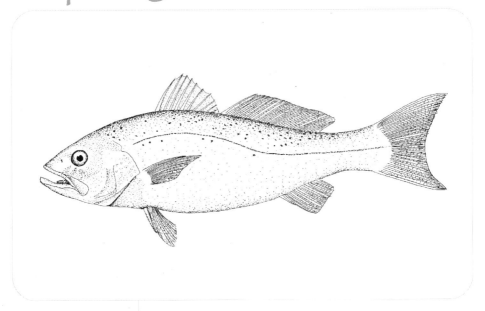

Distinguishing Features and Behaviors

The squeteague, or weakfish, is one of the most easily identifiable fish in Narragansett Bay, and it looks similar to bluefish in fin shape and color. Its anal fin is short, and of the two dorsal fins present, the first is higher than the second. The body is streamlined and slightly flattened, resembling freshwater trout in shape. Its snout is slightly pointed, with a fairly large mouth. Squeteague have two large canine teeth in the upper jaw, and the lower jaw protrudes slightly. The squeteague got the name "weakfish" because of the ease with which a hook tears from its mouth.

Squeteague move near the surface in schools of hundreds. They are fast swimming and active predators, feeding on schools of menhaden, mummi-chogs, other small fish, crabs, mollusks and worms. Juvenile squeteague, in turn, are preyed upon by bluefish and striped bass.

Squeteague belong to a family of fish called drums. Most male drums can produce a croaking or drumming sound, an important behavioral signal used during spawning. They create this sound by contracting abdominal muscles against the swim bladder. Most species of drum have a sensory chin barbel used for bottom feeding, but this barbel is absent from the squeteague.

The northern kingfish *(Menticirrhus saxatilis)*, a species of drum similar to the squeteague, is also common in Narragansett Bay. The kingfish is usually smaller, with dark bars on the side, a longer dorsal fin and a barbel on the chin. Unlike the squeteague, kingfish do not make drumming sounds.

Relationship to People

The squeteague is an important food fish caught both commercially and recreationally. These fish spawn in the quiet coves of Narragansett Bay; however, the population is down, and the observance of young squeteague is cyclical. Few appeared in the Bay during the 1980s, resulting in restrictions being placed on the fishery.

The low population in the Bay is believed to correspond to increased fishing pressure farther south. Juvenile squeteague make up a significant bycatch in the southern shrimp-trawl fishery, resulting in a coastwide population decrease.

FIELD MARKINGS:
Dark olive green above, paler below; back and sides appear iridescent with hints of purple, lavender, green, gold or copper. Sides are marked with small black, dark green or bronze spots above the lateral line. Size: 14 to 26 inches long.

HABITAT:
Shallow waters, open water, along sandy shores, salt marsh creeks.

SEASONAL APPEARANCE:
April to October.

SENSITIVITY LEVEL:

Summer Flounder

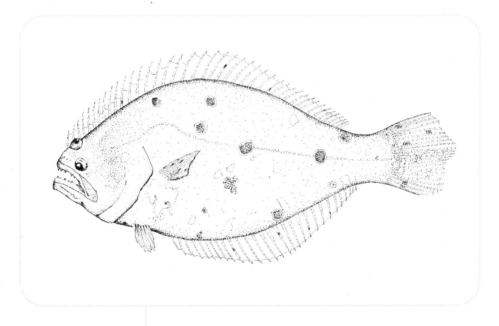

COLLOQUIAL NICKNAMES:
Fluke, "Left-eyed" Flounder

SCIENTIFIC NAME:
Paralichthys dentatus

Distinguishing Features and Behaviors

The summer flounder is the largest species of flatfish in Narragansett Bay. The left side of the summer flounder, or fluke, is scaled and colored and contains both eyes. Its right side, the blind side, has no eyes or scales and is white or translucent in color. This "left-eyed" flounder can be distinguished from other Bay flounder by its large mouth, pronounced jaw, sharp teeth, two identical, narrow ventral fins and at least five dark spots arranged in an "X" on its back.

Summer flounder spend most of their lives close to or on the Bay floor. These fish scatter quickly when disturbed, propelling themselves out of their sandy hiding spots. They are fierce, aggressive hunters and, to ambush prey, they camouflage themselves by flipping sand over their backs. Summer flounder feed on fish, squid, crabs, shrimp, mollusks and worms.

They are closely related to another Bay flounder, called the fourspot flounder *(Paralichthys oblongus)*, which differs in size and color. The fourspot is smaller and thinner with four distinct dark spots on the back. Fourspot flounder can occasionally be seen in summer and fall.

Flounder larvae look remarkably like most other developing finfish, swimming upright in the planktonic zone with one eye on each side of their bodies. During the early life stages of development, the summer flounder's right eye migrates to the other side of the head next to its left eye, while the mouth stays in the original position. After this eye migration, or metamorphosis, takes place, the larvae settles on the bottom.

Fluke spawn in the spring or fall seasons, but never in winter. They spawn along the Atlantic continental shelf, and the young migrate into coastal estuaries such as Narragansett Bay after their first year.

Relationship to People

The summer flounder fishery is one of the most important commercial and recreational fisheries in southern New England, with the commercial value of this species exceeding that of any other fish species. The summer flounder population has fluctuated but appears to be rebuilding.

FIELD MARKINGS:
Changes with bottom color, ranges from all shades of brown to gray-green to almost black. Underside is generally white, although spotting of pigment can occur. Size: 14 to 37 inches long.

HABITAT:
Shallow coastal waters with sandy or muddy bottom; eelgrass and dock pilings.

SEASONAL APPEARANCE:
May to November.

SENSITIVITY LEVEL:

Windowpane Flounder

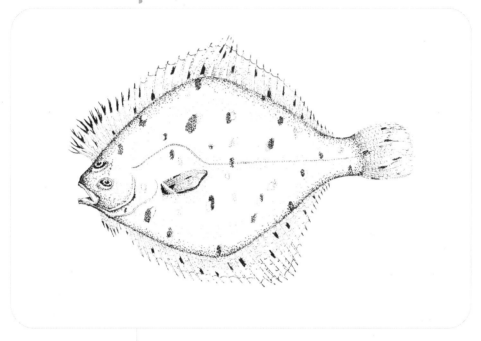

COLLOQUIAL NICKNAMES:
*Sand Dab, Sand Flounder,
Papermouth*

SCIENTIFIC NAME:
Scopthalmus aquosus

Distinguishing Features and Behavior

The windowpane flounder is a common flatfish in Narragansett Bay. Like the summer flounder, this "left-eyed" flounder has both eyes on the left side of its body, and its mouth points to the left.

The shape of the windowpane sets it apart from the other species of flatfish found in the Bay. It is characteristically round in shape, while other flatfish are more oval. The lateral line, used to sense the surrounding environment, is highly arched in this species. Windowpane flounder are thin-bodied, with less muscle than most other flatfish. When held up to a light, it is possible to see through this fish, viewing the outline of its stomach and skeleton, thus the name "windowpane."

The mouth of the windowpane opens wide, and it is easy to see why some refer to it as "papermouth." They have no teeth and feed on shrimp, crabs, small fish and seaweed.

Relationship to People

Historically, windowpane flounder were harvested for use as lobster bait or they were ground into fish meal.

Windowpane flounder are edible but are not as readily sought after as the more thick-bodied summer and winter flounder of Narragansett Bay. As stock sizes of the more desirable species of flatfish have decreased in the Bay, commercial fishing of this species has increased all along the Atlantic coast, resulting in a significant population decline.

FIELD MARKINGS:
Left side is pale, translucent olive green to light brown, with many small, irregular-shaped dark spots and some white spots along the length of the body and on the fins.
Size: 10 to 12 inches long.

HABITAT:
Sandy and muddy bottoms in open water of the Bay.

SEASONAL APPEARANCE:
Year-round.

SENSITIVITY LEVEL:

Winter Flounder

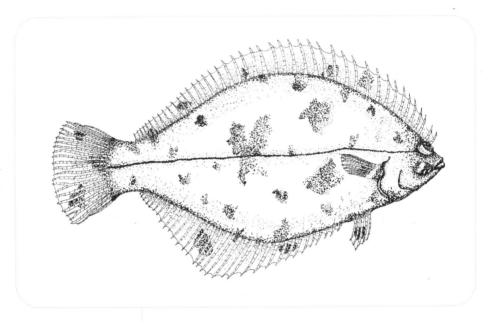

COLLOQUIAL NICKNAMES:
Blackback Flounder,
Georges Bank Flounder,
"Right-eyed" Flounder,
Flatfish,

SCIENTIFIC NAME:
Pseudopleuronectes
americanus

Distinguishing Features and Behaviors

The body of the winter flounder is oval-shaped, flat and thick. Beginning life with eyes on either side of its head, the left eye migrates to the right side within a few weeks, thus earning the name, "right-eyed" flounder. The eyes remain close together on the upper side of the winter flounder. The left, or blind side, of the fish is white and faces the bottom.

Unlike other Bay flounder, winter flounder have rough scales and a small mouth, with thick, puckered lips and small rows of slightly rough, flattened teeth. These fish can change color to blend in with the bottom, but are generally much darker than most other Bay flatfish.

Flatfish

Narragansett Bay and nearby coastal salt ponds are important spawning areas for winter flounder. Mature adults migrate from deeper waters in the Bay and Rhode Island Sound into shallower waters during late fall. Spawning occurs from late December into April. They deposit clusters of sinking eggs in slow-flowing coves and embayments. Juvenile fish remain in shallow nursery areas for two to three years before migrating to deeper water. Winter flounder are the only Bay flatfish that spend their entire early life in the estuary.

Winter flounder prefer sandy or muddy bottoms and are sometimes found near eelgrass beds, but can also be found on many other bottom types. They are omnivores and feed on shrimp, crustaceans, amphipods, larval fish, mollusks, worms and some types of seaweeds and plants.

Relationship to People

Winter flounder were once one of the more important recreational and commercial fish species in Narragansett Bay. Because the winter flounder spends the majority of its life in estuarine environments, this species is affected by both habitat degradation and overfishing.

The winter flounder population decreased dramatically in the late 1980s and early 1990s. In 1992, the catch numbers reached an all-time low. A Bay-wide ban on fishing winter flounder went into effect in 1991. Although some fishing has been permitted, strict regulations will continue until local stocks recover. Warming winter water temperatures have increased sand shrimp predation of juvenile winter flounder, which may explain some of the decline of winter flounder populations in Narragansett Bay.

FIELD MARKINGS:
Adults vary in color from shades of dark brown to gray or olive green; may have mottled blotches and light specks. Juveniles are lighter and have more spotting. Size: averages 12 inches long.

HABITAT:
Open water, muddy and sandy bottom.

SEASONAL APPEARANCE:
Year-round; most common in winter for spawning.

SENSITIVITY LEVEL:

American Eel

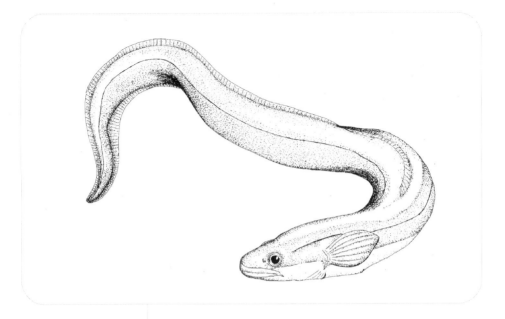

SCIENTIFIC NAME:
Anguilla rostrata

Distinguishing Features and Behaviors

American eels are long, snakelike fish common along the Atlantic coast. The eel's body is rounded, with a long jaw, pointed snout and two nostrils in front of its small eyes. The dorsal, caudal and tail fins are joined and continuous from the middle of the back, around the tail, to the middle of the belly. It has no pelvic fins, and its scales are small in size and hardly noticeable. Often misidentified as sea snakes, eels are distinguishable by the continuous fin that is absent in snakes.

As a catadromous fish, American eels migrate from fresh water to spawn in saltwater environments. Adults, ages five to 35, migrate from freshwater creeks and rivers to spawn in the warm, nutrient-rich waters of the Sargasso Sea, an area in the Caribbean, east of the

Bahamas and north of the West Indies. The adults die after spawning, and the larvae spend the next year drifting and swimming with the Gulf Stream to northern coastal waters. The European eel, *Anguilla anguilla*, is a similar species that migrates a great distance to spawn in the Sargasso Sea. They drift farther north in the Gulf Stream to Europe, while the American eel larvae move west once they reach the Atlantic coast.

American eel larvae do not resemble adults at hatching; instead, they are leaf-shaped, long, transparent and flattened sideways. After the "leptocephalus" larval stage, the eel transforms during its journey to fresh water, finally resembling a small adult by the time it arrives. Most of the young eels, or elvers, migrate toward freshwater streams and rivers, while some remain in the muddy waters of tidal creeks and marshes. Females will migrate farther into fresh water than males.

Eels are nocturnal scavengers, preferring the night environment to forage for food. They are omnivorous, feeding on detritus, small fish and invertebrates.

Relationship to People

American eels are often used as live bait in the recreational striped bass fishery of Narragansett Bay. Eels eagerly bite on most bait and are generally caught at night. Fish ladders built for herring also benefit eels. During spring nights, elvers can be observed in large numbers as they make their way to fresh water. Eels were recently considered for listing as an endangered species, and while they did not make the list, strict conservation and management is warranted.

American eels are a prized food source in European and Asian cuisine, often served in sushi.

Field Markings:
Greenish-brown above, fading to yellowish below Size: up to 4 feet long. Females are larger than males.

Habitat:
Muddy bottoms of freshwater rivers, tidal creeks, harbors, salt ponds.

Seasonal Appearance:
Spring to fall; buries in mud during winter months.

Sensitivity Level:

Atlantic Menhaden

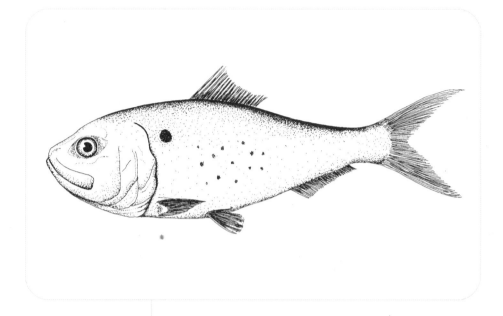

Distinguishing Features and Behaviors

Menhaden are members of the herring family, with a thick body flattened sideways and a sharp-edged belly. The menhaden's large head — almost one-third of its total body length — distinguishes it from other fish in the herring family, such as alewives and shad.

The menhaden makes extensive seasonal migrations, traveling in schools of hundreds or even thousands. Fish in these schools swim in unison, following a single lead fish. When near the shore, they often break the surface with their fins and tails.

Although menhaden spawn in the ocean, eggs, larvae and juveniles can be found in the Bay during the summer. They are not anadromous like similar herring species that live in salt water and spawn in fresh water.

Instead, menhaden migrate into the Bay to feed on the rich supply of plankton present in the summer and fall.

The menhaden feeds by opening its mouth and allowing water to pass through its gill openings, which filter microscopic plants and small crustaceans from the water. In calm water, the snouts of menhaden often emerge as they feed on the surface-dwelling plankton. Menhaden are preyed upon by a wide range of predators, including whales, porpoises, bluefish, striped bass, cod, swordfish, tuna and birds.

Relationship to People

The menhaden, which is fatty, is processed commercially as chicken and pet feed, fertilizer, oil for plants, soap, ink, cosmetics and tampering products for steel. Because menhaden is so oily, it is not harvested for food.

Menhaden are considered excellent live bait for bluefish. Fishermen will set on schools of feeding menhaden, using hooks to snag individuals. The snagged fish are used as live bait to catch large striped bass that feed upon the menhaden.

Historically, menhaden supported one of the oldest and largest fisheries on the East Coast. They were harvested for agricultural fertilizer during the 1600s, and later to support an oil-extraction fishery. Presently, menhaden fishers in Narragansett Bay sell most of their catch as bait for lobster traps or for recreational fishing. Populations of menhaden have declined since the 1970s due to overfishing. Coastwide, the Rhode Island the menhaden population has begun to recover but will need strict management to continue this encouraging trend.

Menhaden are harvested for "reduction" or industrial uses in Virginia; that practice was banned in Rhode Island in 2003.

FIELD MARKINGS:
Silvery with lustrous, brassy sides and a dark blue-green back. Adults have numerous spots on their sides, located behind a larger, dark shoulder spot. Size: 12 -15 inches long.

HABITAT:
Seagrass meadows, open water.

SEASONAL APPEARANCE:
Spring, summer, fall.

SENSITIVITY LEVEL:

Atlantic Salmon

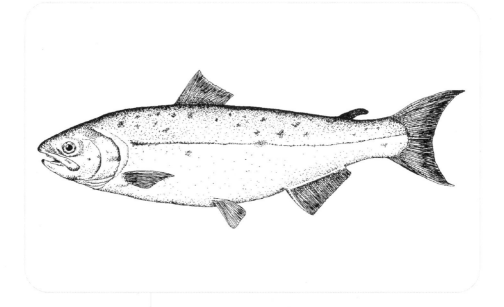

COLLOQUIAL NICKNAMES:
Sea Salmon, Black Salmon

SCIENTIFIC NAME:
Salmo salar

Distinguishing Features and Behavior

The Atlantic salmon is a slender and graceful fish whose Latin name means "the leaper." Its distinctive characteristics make the Atlantic salmon easy to recognize: a small head, blunt nose, small eyes and a mouth that gapes back below its eyes. The mouth contains a row of stout, conical teeth. Salmon have large scales and a slightly forked caudal fin. One distinguishing characteristic of the Atlantic salmon is the presence of an adipose fin, a feature in all species of trout.

The Atlantic salmon is an anadromous fish that lives in salt water but migrates to freshwater rivers in the spring to spawn. This migration activity is called a run. Salmon have been known to swim upstream in larger

rivers for up to 200 miles in order to reach their native
breeding grounds.

The Atlantic salmon is quite healthy and silvery when it
enters the river to spawn, but after spawning, it weakens
and becomes dull in color. Large black spots develop on its
skin, the fins thicken, its jaws elongate and its skin becomes
covered in slime. Unlike other salmon that die immediately
after spawning, many Atlantic salmon return to the ocean
afterwards. These rugged survivors are called kelts.

Young salmon migrate downstream to the ocean at two
years of age, where they feed and grow. Their diet consists
of crustaceans and small finfish. When the salmon reaches
adulthood at about three years, it becomes a voracious
predator, feeding only on large prey. Atlantic salmon
are usually so strong that only large fish, such as tuna,
swordfish or sharks, are able to eat them.

Relationship to People

The Atlantic salmon was abundant in New England rivers
until the late 1700s when they began to disappear. The
Industrial Revolution brought dams, deforestation, habitat
alteration and pollution, While all of these factors resulted
in the loss and obstruction of spawning rivers and habitat,
dams have been the primary cause of the population
decline.

Most of the known freshwater salmon runs in New
England have been eliminated due to loss of habitat. The
only known naturally occurring runs in the U.S. are found
in Maine, with fewer salmon returning to spawn in these
rivers each year. Rhode Island has a salmon restoration
program that aims to restore natural spawning runs, but
the wild population in Narragansett Bay is extirpated.

FIELD MARKINGS:
Silvery body with
some dark crosses
and spots on the
head, body and fins.
Males have red
patches along the
sides. In fresh water,
juveniles have 10
or 11 dark crossbars
alternating with
bright red spots.
Size: 2 to 3 feet
long; up to 10
pounds.

HABITAT:
Freshwater rivers,
streams, lakes, open
bays, coastal waters.

SEASONAL APPEARANCE:
Spring and summer.

SENSITIVITY LEVEL:

Eastern Brook Trout

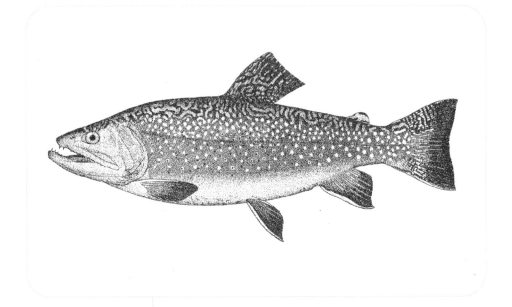

Distinguishing Features and Behavior

The brook trout is the only native member of the salmon family that still breeds in the wild within the Bay watershed. Brown and rainbow trout and salmon are all hatchery-raised.

COLLOQUIAL NICKNAMES:
Sea Trout

SCIENTIFIC NAME:
Salvelinus fontinalus

The brook trout's body is silvery-green with yellow to reddish-brown spots and a pale brown to red underside. They range in size from six to nine inches in streams, but the sea-run variety (called 'salters') may grow up to 25 inches and weigh more than four pounds.

The eastern brook trout is known to spawn in cold freshwater rivers and streams. While sea-run varieties have been documented around the New England coast, it is unclear whether any remain in Narragansett Bay. Brook trout have declined precipitously in this region over the past several decades, primarily as a result of habitat destruction and pollution. They require cold, clean water and are sensitive to impacts from development such as sedimentation and loss of wetlands.

Relationship to People

Brook trout are a prized game fish species and can be caught in small streams where they may only grow to a few inches. They are good to eat, but there is no market or commercial fishery for them due to their typically small size and remote habitat.

Brook trout are important indicators of river water quality. Efforts to restore native brook trout habitat through land conservation and stormwater control must become a high priority if we are to reverse the population decline.

FIELD MARKINGS:
Silvery-green with yellow to reddish-brown spots and pale brown to red underside. Size: about 6 to 9 inches in streams, but the sea-run variety may grow up to 25 inches and weigh more than 4 pounds.

HABITAT:
Freshwater streams and rivers, but may migrate into the estuary and ocean.

SEASONAL APPEARANCE:
Year-round.

SENSITIVITY LEVEL:

River Herring

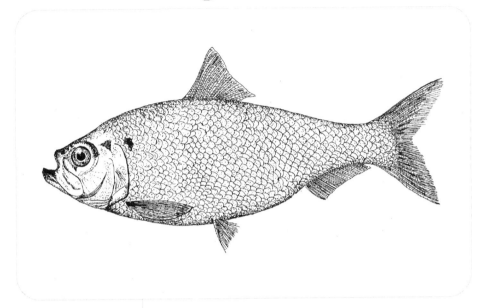

COLLOQUIAL NICKNAMES:
Sawbelly, Bucky, Gaspereau,
Blueback, Glut Herring,
Big-eye

SCIENTIFIC NAME:
Alewife: Alosa
pseudoharengus,
Blueback Herring:
Alosa aestivalis

Distinguishing Features and Behaviors

River herring is the general name for two species of fish commonly found in the Bay: the alewife and the blueback herring. They are similar to one another, differing only slightly in appearance and behavior. The blueback herring is thinner while alewives are thicker and more greenish-black. This distinction is most apparent in freshly caught fish. River herring are also similar to shad.

River herring are thick around the head and abdomen, narrowing to the tail with heavy sharp serrations along the underside, hence the nickname "sawbelly." Its scales are large and slough off easily when handled. It has a distinctive dusky gray spot just behind the margin of the gill cover. This fish has large eyes, a lower jaw that projects beyond the upper jaw and a toothless mouth.

River herring are anadromous fish that spend their adult lives in the ocean and return each year — congregating in schools of thousands — to spawn in the freshwater streams where they hatched. River herring suffer high rates of natural mortality, with fewer than 1% of all eggs surviving the harsh migration to salt water as juveniles. Most herring die during or after spawning migrations.

The river herring feeds primarily on plankton, including copepods, amphipods, shrimp and fish eggs. It is preyed upon by striped bass, bluefish, gulls, terns and other coastal birds.

Relationship to People

Caught mainly in the mouths of streams and rivers during spawning, both species of river herring are commercially important as a food source and as bait. The river herring's habitat range has been largely restricted due to overfishing, pollution created by manufacturing waste and the erection of dams. These dams — many of which are still in existence today — block rivers vital to spawning migrations and greatly reduce the ability of the population to be sustained.

Once a rite of spring in Rhode Island, fishing for river herring was banned in 2005 after a 95% decline was observed between 2000 and 2005. No one knows why river herring declined so dramatically, but theories include climate change, overfishing and increased predation by striped bass.

State and nonprofit environmental groups are attempting to reestablish alewife stocks by building fish ladders over dams, cleaning up rivers and introducing spawning adults from other locations to enhance the existing Rhode Island alewife population.

FIELD MARKINGS:
Pale white and silvery with a light gray-green back. Appears iridescent with shades of green and violet in the water. Single dark spot on the shoulder just behind the gill cover. Size: up to 15 inches long. Females are larger than males.

HABITAT:
Salt marshes, open water, freshwater rivers, river mouths.

SEASONAL APPEARANCE:
Spring through late fall.

SENSITIVITY LEVEL:

Striped Bass

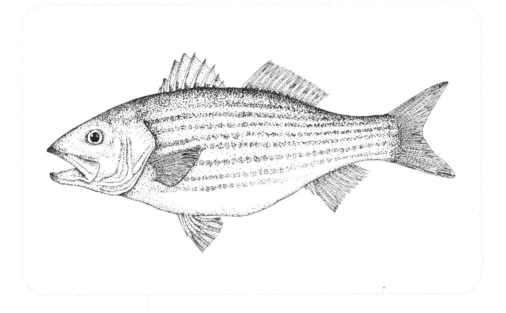

COLLOQUIAL NICKNAMES:
Striper, Rockfish

SCIENTIFIC NAME:
Morone saxatilis

Distinguishing Features and Behaviors

One of the most widely recognized fish in Narragansett Bay is the striped bass. The body of the striped bass is thick and stout and is lined with seven to eight narrow horizontal stripes, the highest being its most distinctive. The striped bass has two well-developed dorsal fins, one spiny and one soft-rayed, plus a wide, forked tail. The mouth is large with small teeth, and its lower jaw protrudes slightly.

This fish is a powerful swimmer, able to swim in harsh surf environments. Most striped bass travel in large schools, except for the very large fish, which travel solo. For the first two years of their lives, striped bass live in small groups and are often called "schoolies." Female fish grow larger than males and are referred to as "cows." Most bass longer than 30 inches are females.

Striped bass feed on many species of finfish, including alewives, menhaden, flounders and silversides, as well as many invertebrates, such as lobsters, crabs, clams, squid and worms.

The striped bass is an anadromous fish, meaning it migrates from salt water to fresh water for spawning. These fish undertake long migrations down the Atlantic coast to spawn in the Chesapeake Bay and Hudson River estuaries each spring and migrate up the coast during warmer summer months. It is during these northern summer migrations, after spawning, that they appear in the Bay. Many of the individuals seen seasonally in New England waters are believed to have originated in the Hudson River.

Relationship to People

The striped bass is one of the most highly prized saltwater game fish in Narragansett Bay. Most stripers are caught by recreational anglers, but many are caught by commercial rod and reel as well. Stripers are a delicious food fish that reproduce well in hatcheries and are marketed commercially.

The striped bass population in New England has undergone periods of abundance and decline due to overfishing and degraded spawning habitats. In 1986, a moratorium was placed on all striped bass fishing in Rhode Island, Connecticut and New York, due to evidence of a sharp decline in the population. This moratorium led to the creation of strict management measures, including minimum sizes and quotas. Striper populations have recovered dramatically from the Chesapeake to the Canadian Maritimes.

FIELD MARKINGS:
Dark olive green to dark blue on the tip, with silvery-gray sides and white belly; seven to eight dark horizontal stripes. Size: averages 20 to 30 inches long and about 5 pounds. Can grow up to 5 feet and weigh 70 to 80 pounds.

HABITAT:
Open water along rocky shores, sandy beaches, salt ponds, rivers.

SEASONAL APPEARANCE:
Early April to late fall; migrate south during winter.

SENSITIVITY LEVEL:

Northern Puffer

COLLOQUIAL NICKNAMES:
Puffer, Swellfish,
Swell Toad, Toad,
Balloonfish, Blowfish

SCIENTIFIC NAME:
Sphoeroides maculatus

Distinguishing Features and Behaviors

Pufferfish are a commonly known family of fish named for their ability to inflate themselves by swallowing water (or air, if removed from water). When this occurs, they become round and softball-sized, causing them to appear larger to predators. Striped bass and bluefish are the puffer's predators in Narragansett Bay.

The mouth of this fish is small, with teeth fused into powerful cutting and crushing beaks divided in the middle. The northern puffer has sandpapery and teethlike scales, with small, crescent-shaped gills in front of its pectoral fins. Puffers sometimes target their prey in large schools. Their diet consists of large crabs, other hard-shelled invertebrates and some finfish.

The northern puffer spawns May through August off of the Massachusetts coast, laying sticky, transparent eggs in bottom areas. It is thought that their larvae may be dispersed throughout the North Atlantic coast by the Gulf Stream current. The northern puffer's population is highly resilient, with a quick reproductive rate.

Around the Bay, the northern puffer lives in a variety of habitats. They have been collected while seining through shallow brackish estuaries, by trawling in saltier areas of the southern Bay and at a depth range to nearly 600 feet. They are usually only seen as younger fish and aren't found in the colder months. Their lowest temperature limit is about 46°F.

Other species of puffers found in Narragansett Bay during the early fall months include the bandtail pufferfish *(Sphoeroides spengleri)*, which has a row of large dark spots along the lower part of its body from chin to caudal-fin base, and the striped burrfish *(Chilomycterus schoepfi)*, which has dark brown dorsal stripes with rigid spiny skin that is skeletonlike.

Relationship to People

Some puffers produce a powerful poison (tetrodotoxin) that is concentrated in their major organs, including their skin. Northern puffers are reportedly non-toxic, and are sold as "sea squab" in the northern part of their range. Puffers are also a popular aquarium fish, especially the porcupinefish and burrfish. They are commercially important fish for both of these reasons.

FIELD MARKINGS:
Upper side is gray or brown with tiny black pepper-spots. Belly is yellow to white. Black bar between eyes is often present. Size: maximum 14 inches.

HABITAT:
Bays, estuaries and protected coastal waters around rocks or reefs.

SEASONAL APPEARANCE:
Late spring through fall.

SENSITIVITY LEVEL:

Seahorse and Pipefish

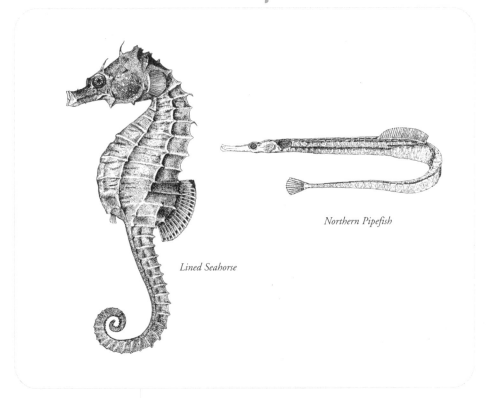

Northern Pipefish

Lined Seahorse

COLLOQUIAL NICKNAMES:
Lined Seahorse, Northern
Seahorse, Spotted Seahorse
and Northern Pipefish

SCIENTIFIC NAME:
Seahorse:
Hippocampus erectus
Pipefish: Syngnathus fuscus

Distinguishing Features and Behaviors

Pipefishes and seahorses belong to the family Syngnathidae, which translates from Greek to "with jaw," emphasizing their long snouts. The seahorse's body contains a series of ridges running length-wise from head to prehensile tail. Additional support comes from the bony rings that bisect the ridges across its body.

The coloration of the lined seahorse is dependent upon the color of its surroundings and may be dull grayish, olive green, brown, yellowish or brick red. Their bulbous, amorphous bodies make seahorses poor swimmers, despite their fan-shaped dorsal fin and pectoral fins. Its uniquely adapted tail is long and coiled in order to secure the animal to blades of seagrass, seaweed, sargassum or coral that are anchored on the bottom. Pipefishes differ in that their long, straight tails possess a caudal fin. The northern pipefish may enter freshwater ecosystems.

Seahorses undertake a lengthy courtship process involving clicks and other sound-making. They are among the few animal species to practice monogamy, remaining with their mate for life. The male broods the developing off-spring in his pouch for two to three weeks. Northern pipefish reproduce similarly; males carry eggs in brood pouches located on the ventral side of their tails.

Since they are very slow swimmers, lined seahorses rely on camouflage to hide them from larger fish and other predators. Additionally, their feeding strategy must compensate for their inability to move quickly after prey. They draw in vast quantities of water with their elongated horselike mouth and their diet consists of small crustaceans such as brine shrimp and marine worms. Pipefish eat zooplankton, brine shrimp and other benthic invertebrates.

Relationship to People

In some cultures, the seahorse is considered an aphrodisiac and medicinally beneficial (for asthma, throat problems, infertility and other health issues). Millions also are collected annually from their native habitats around the world for aquarium exhibits. The survival of Narragansett Bay's lined seahorses and northern pipefish is dependent upon maintaining healthy eelgrass bed habitats.

FIELD MARKINGS:
The lined seahorse's body can sometimes appear mottled and occasionally shows small light-colored spots. Size: averages 6 inches or less; up to 7-8 inches long.

The northern pipefish's body is very dark and lacks a distinctive coloration pattern. Size: up to 1 foot long.

HABITAT:
Marine or brackish water with adequate vegetation, eelgrass beds, coral reefs.

SEASONAL APPEARANCE:
Spring through fall (migrate toward the shelf in winter).

SENSITIVITY LEVEL:

Tropical Migrants

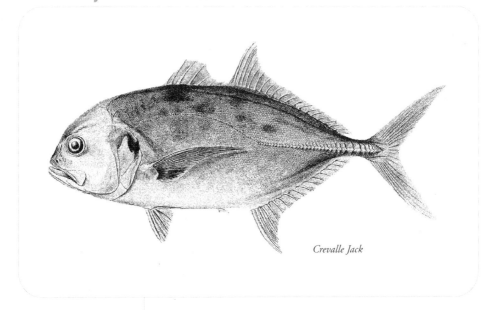

Crevalle Jack

COLLOQUIAL NICKNAMES:
Crevalle

SCIENTIFIC NAME:
Caranx hippos

Tropical and subtropical originating juvenile fish are found in Narragansett Bay in the late summer and early fall when the water temperature is very warm and similar to that of their natural range. They live on muddy bottoms in brackish estuaries or other coastal marine waters — ideal because they act as nursery grounds, so there is plenty of small prey. Most of these transient fish feed on smaller fish, worms and crustaceans such as small crabs and shrimp.

The adults of these tropical species spawn offshore, south of Cape Hatteras, North Carolina. Eggs and larvae are transported northward to temperate regions by the Gulf Stream current. Despite being exotic, many of these species have been caught all over the western North Atlantic and have been observed by snorkelers and divers in shallow and rocky areas.

Due to their bright colors and patterns, many of the transient fish that arrive in Narragansett Bay are used as aquarium fish. Their status as native to the tropics and sub-tropics makes local sightings notable each summer season.

Here, we highlight two of the more commonly seen tropical migrants.

CREVALLE JACK

Distinguishing Features and Behaviors

Crevalles have a deep body and a steep head profile. As members of the jack family, crevalles are fast, graceful swimmers with deeply forked tails.

Small juveniles travel in fast-moving schools but prefer to be on their own as they grow larger. Working together, crevalles can corner or herd baitfish into small bait balls where they are voracious predators. They feed on smaller fish, shrimp and other invertebrates. When caught, they sometimes make a grunting or croaking noise by grinding their teeth together while releasing gas from their air bladder. Crevalles conduct this action in the hope of deterring predation.

Relationship to People

Within their usual range, crevalle jacks are an important species in commercial fisheries. They are also a valued game fish because of their quick, strong maneuverability. Fishermen come in contact with crevalles when targeting other species offshore where the crevalle jack is prey for surface-feeding carnivores such as tunas, marlins and shorebirds.

FIELD MARKINGS:
Silvery-blue to brassy on top and sides; often yellow on underside and fins Size: up to 5 feet and 20 pounds.

HABITAT:
Brackish estuarine waters with muddy or sandy bottoms, seagrass beds.

SEASONAL APPEARANCE:
Late summer through early fall.

SENSITIVITY LEVEL:

Tropical Migrants

Spotfin Butterflyfish

COLLOQUIAL NICKNAMES:
Spotfin Butterflyfish,
Two-spotted Butterflyfish,
Banded Butterflyfish

SCIENTIFIC NAME:
Chaetodon ocellatus,
Chaetodon striatus

SPOTFIN AND BANDED BUTTERFLYFISH

Distinguishing Features and Behaviors

Butterflyfish tend to be disc-shaped and very laterally compressed. Juveniles have many rigid, bony head plates, and both juveniles and adults are brightly colored. A vertical black bar passes from the forehead through the eye to the anal fin. Juveniles possess a second black bar parallel to that of the adult form. The spotfin's common name is derived from the dark black speck at the upper tip of its soft dorsal fin. The banded butterflyfish is white with three vertical black body stripes and one through its eye.

The majority of the 114 species of butterflyfish breed in monogamous pairs. They are territorial and often remain within the same general vicinity of where they first settled as benthic juveniles.

Spotfin and banded butterflyfishes belong to the family Chaetodontidae, which means "bristle-toothed" in Greek and describes the presence of rows of rough-textured teeth. Each fish has a small protruding snout used to search the reef's cracks and crevices for anemones, worms, coral polyps and other invertebrates. Banded butterflyfish occasionally clean other fish or feed en masse on plankton.

Relationship to People

A popular aquarium fish, spotfin butterflyfish are traded commercially. Banded butterflyfish are more resistant to captivity because of their unique diets. Both species have high fecundity and low population doubling times, which increase their resilience to human collecting pressures.

Increasing water temperatures resulting from human-induced climate change may increase larval transport and potential survival rates in warming mid-Atlantic and New England waters.

Some **Other Tropical Migrants** commonly seen in Narragansett Bay are: Banded rudderfish *(Seriola zonata)*, bluespotted coronetfish *(Fistularia tabacaria)*, bicolor damselfish *(Pomacentrus partitus)*, inshore lizardfish *(Synodus foetens)*, planehead filefish *(Stephanolepis hispidus)*, short bigeye *(Pristigenys alta)* and the snowy grouper *(Epinephelus niveatus)*.

FIELD MARKINGS:
Predominantly whitish-gray with a bright yellow margin extending from dorsal to anal fin and a black spot or bar near the head. Size: up to 8 inches in length. Juveniles range from half an inch to 1 inch. Banded butterflyfish reach a maximum size of about 6 inches.

HABITAT:
Coral and rocky reef environments.

SEASONAL APPEARANCE:
Late summer through early fall.

SENSITIVITY LEVEL:

Because it is so difficult to pinpoint one preferred habitat for birds, there is no quick-reference icon in this section. The various places you might find a particular species are still noted within the text.

Birds

Narragansett Bay is the stage for very diverse bird life. The Bay's hundreds of small embayments, islands and salt marshes provide excellent habitat for waterfowl, sea birds and shorebirds. Of the documented 155 bird species inhabiting Rhode Island, roughly 40 of these rely on the uninhabited islands and coastal wetlands of Narragansett Bay for breeding grounds and migratory refuges.

The *Uncommon Guide* gives concise identification keys and a brief local history for 31 birds. These particular species have been included for their commoness to Rhode Island waters and for their importance in the Bay ecosystem. They can be divided into six general categories: waterfowl, waders, shorebirds, gulls and terns, birds of prey and other.

Waterfowl — ducks, geese, eiders, loons and cormorants — are predominantly swimming birds that feed on eelgrass, seaweeds, fish and crustaceans. Throughout fall and winter, waterfowl such as common goldeneye and bufflehead rely on the sheltered waters and the abundant food of the Bay.

Wading birds comprise egrets, herons and ibises. These birds walk along the shallow areas around salt marshes and estuaries, hunting for fish and invertebrates. Shorebirds include sandpipers, plovers and others found running along the water's edge, hunting for small insects and crustaceans.

Gulls and terns are a familiar site around the Bay. They are one of the most commonly seen bird species at any time of the year. Birds of prey include the northern harrier and the osprey. Northern harrier cruise over salt marshes and upland fields, hunting for mice and voles, while osprey feed mainly on fish.

This section of the *Uncommon Guide* is dedicated to the birds commonly seen on and around Narragansett Bay. It offers illustrations and field markings to help the user identify various species. In most cases, field markings are specific to the plumage appropriate for the season during which a bird is commonly seen. For example, the common loon's eclipse plumage is described since loons are typically seen in Rhode Island during winter.

In addition, this section offers the user a brief introduction of the ecology of each species of bird, including brief natural histories of these birds and their relationship to people and to Narragansett Bay.

American Black Duck

Distinguishing Features and Behaviors

A large dabbling duck, the American black duck is often confused with the mallard hen. Black ducks have brilliant violet wing patches bordered with white. Distinct white wing linings are noticeable in flight and differentiate black ducks from mallards.

Scientific Name:
Anas rubripes

Black ducks dredge aquatic invertebrates, such as mussels, quahogs and periwinkles, out of the sediment. They use their broad bills to hold and open shellfish. It is common to see black ducks dabbling or splashing around as they feed on vegetation with their tails tipped upright. The color of the bill is used to distinguish the male and female of the species: the male's bill is yellow and the female's is olive green with dark patches.

Black ducks nest on the ground in wetlands where vegetation supplies camouflage and nesting material. American black duck pairs breed during the spring, laying eight to ten eggs that hatch after 26 to 29 days of incubation. The young are born with a downy coat and are mobile one to three hours after hatching. While salt marshes may be used for nesting, black ducks rely heavily on coastal habitats during the winter when freshwater areas are more likely to be iced over.

Relationship to People

The American black duck was once the most abundant breeding duck in the United States; however, it was placed on the Audubon Society's Blue List from 1980 to 1981 for species that have undergone population reductions. The population was decreasing due to aerial spraying of pesticides, loss of habitat and the increasing natural hybridization with mallards. It is common to see mallard/black duck hybrids.

Black ducks are a favorite among New England duck hunters because of their tendency to inhabit inland, accessible waters. Due to its dabbling manner of feeding, this species was once prone to fatally ingesting lead pellets from hunters. The U.S. Fish and Wildlife banned the use of the lead shot for hunting waterfowl in 1991. Since then, studies have shown a significant reduction in black duck mortality; the population is reported to have stabilized in the last 20 years.

The black duck is a year-round resident and breeder. A recent bird survey indicates 28 confirmed American black duck breeding sites, with an additional 32 possible or probable in Narragansett Bay.

FIELD MARKINGS:
Mostly dark brown with paler head and neck. Violet wing patches bordered by white. The bill is broad, yellow on the male and olive with dark patches on the female. Size: 22 inches long.

HABITAT:
Salt marshes, brackish and fresh water ponds.

SEASONAL APPEARANCE:
Year-round.

SENSITIVITY LEVEL:

American Wigeon

Distinguishing Features and Behaviors

The American wigeon is a medium-sized dabbling duck often found interacting with other dabbling ducks, such as the American black duck and the mallard. Occasionally, they are found near diving ducks like the bufflehead and the common goldeneye.

The male, or drake, is identified by his white crown, earning the nickname "baldpate." In flight, wigeons have a large white patch on the forewing. The males of most North American species of waterfowl, including the America wigeon, molt their feathers after mating season, exchanging colorful breeding plumage for a dull, drab appearance referred to as the eclipse plumage. In this state, the drake appears similar to the wigeon hen because the drake loses the white feathers of the crown.

The American wigeon feeds primarily on aquatic vegetation including microalgae. They graze on tubers, seeds and stalks by dabbling with their tail upright out of the water. These ducks are notorious for stealing vegetation from other diving ducks, including scaup, redheads and mergansers.

Relationship to People

This migratory bird has recently begun breeding on the Atlantic coast; however, it has not established breeding grounds in Narragansett Bay. Despite the absence of breeding grounds, the Bay hosts the largest transient wigeon population in New England. Over 1,200 wigeons have been observed roosting in Narragansett Bay and Rhode Island waters during their fall migration. Several hundred wigeon winter in the Bay each year.

Because American wigeons feed on submerged vegetation, they are susceptible to ingesting lead pellets from gunshots. These pellets dissolve in the digestive system, gradually allowing lead to enter the blood stream, a common cause of death among dabbling ducks.

FIELD MARKINGS:
A medium-sized duck with a light blue bill. The male has a brown head with a green patch and a white crown. The female is mostly brown, with a gray head and neck. Size: 18 to 23 inches long.

HABITAT:
Marshes, inland coastal waters.

SEASONAL APPEARANCE:
Migratory; fall and winter.

SENSITIVITY LEVEL:

Brant Goose

Distinguishing Features and Behaviors

COLLOQUIAL NICKNAMES:
Brant

SCIENTIFIC NAME:
Branta bernicla

The brant goose is a small, dark sea goose. The goose's genus name is derived from the German word "brand" meaning "burnt" and refers to its dark feathers. The species name *bernicla* is from the Norwegian word for "barnacle." An ancient legend associated with brant is that the geese hatch from barnacles on driftwood.

Brant have long and pointed wings, and flocks fly low with rapid wing beats in ragged formation. The flocks do not form a "V" shape as some geese do; instead they bunch together or form long wavy lines.

Brant travel along the coast and are usually found on sandy peninsulas and bars. They generally avoid migrating over land. They return to the same nesting site year after year as many birds are "pre-programmed" to do. Brant are monogamous, forming lifelong pair bonds at three years of age.

Large flocks of brant reside from fall through late spring in the upper Bay and can often been seen feeding close to shore along the Providence River.

Traditionally, the main Bay food source for brant was eelgrass. However, eelgrass decline in the northeast has led biologists to speculate that these geese now feed primarily on macroalgae, along with lesser amounts of aquatic plants, moss, lichen, crustaceans, mollusks, worms, insects and grain. Brant feed in a manner similar to ducks, dipping from the water's surface and dabbling in the submerged vegetation.

To survive in a strictly saltwater environment, brant have adaptations that make them true sea geese. Salt glands located at the base of the bill filter excess salt from their blood stream, concentrate it and excrete it out through the bill. This allows the birds to drink sea water and eat saltwater vegetation without becoming dehydrated.

Relationship to People

The Atlantic brant population was decimated in the 1930s and 1940s when up to 90% of the Atlantic coast's eelgrass beds were killed by disease.

Recent brant population declines in Narragansett Bay have been attributed to a decline in eelgrass. The population has not returned to its former numbers even though brant have adapted to eating sea lettuce and other vegetation.

FIELD MARKINGS:
Dark-colored with a black head, neck and breast. White patch on both sides of the neck, just under the throat. Underside of the body is gray to white at the tail, with a black bar at its end. Juveniles have a bold white edge to the wing feathers, giving them a wave pattern across their back. Size: 25 inches long.

HABITAT:
Salt water inlets, estuaries, coastal shores.

SEASONAL APPEARANCE:
Fall, winter, spring.

SENSITIVITY LEVEL:

229

Bufflehead

Colloquial Nicknames:
Butterball

Scientific Name:
Bucephala albeola

Distinguishing Features and Behaviors

Buffleheads are the smallest species of North American diving ducks. They are commonly seen in inland bays and coastal ponds, preferring open water without vegetation.

These small ducks often congregate in flocks of up to 50 birds; they fish in numbers to increase catch rates. When feeding, one duck will often remain above water as a lookout, watching for predators.

Throughout the year, buffleheads feed on aquatic insects almost exclusively. On their wintering grounds along the Atlantic Coast, they feed on crustaceans and mollusks. In Narragansett Bay, buffleheads feed mainly on sand shrimp and periwinkles.

Like many North American diving ducks, buffleheads use the Bay as a refuge during migration. These birds — one of the few monogamous species of duck — nest and breed in the forests of Canada and Alaska during spring and summer. After breeding season, they scatter southward toward more forgiving climates. Birds that nest in the eastern provinces of Canada commonly migrate to the Atlantic coast in the winter.

Due to the availability of unfrozen inland water and the abundance of aquatic invertebrates and small fish, Narragansett Bay is an important winter staging area for species of ducks such as bufflehead, scaup and goldeneye.

Relationship to People

During the winter months, buffleheads can be seen in sheltered coves and marshes, usually flocking together in groups of ten to 100 individuals. The sheltered water of inland bays and ponds poses a threat for buffleheads in fall and winter, making them vulnerable to hunting.

Even with potential hunting pressure, buffleheads are one of the few species of ducks whose population has increased significantly since the 1950s.

FIELD MARKINGS:
Males are mostly white with a dark back. They have a distinct white patch on a large, puffy head. Females are dark grayish-brown, with a distinct white cheek spot. In flight, both sexes have a large white wing patch. Size: 13 to 15 inches long.

HABITAT:
Offshore, in bays, coastal ponds.

SEASONAL APPEARANCE:
Migratory; late fall, winter, early spring.

SENSITIVITY LEVEL:

Canada Goose

SCIENTIFIC NAME:
Branta canadensis

Distinguishing Features and Behaviors

The Canada goose is one of the most common and well-recognized species of geese, often incorrectly called a Canadian goose. They are heavy-bodied birds with dark wings, a pale breast and white under the tail. Distinguishable in flight by their extended necks, they also have a white "U"-shaped band on their rumps. When migrating, the geese flock together in a "V" or "W" formation, calling almost continuously while flying. Their distinct "honk-a-lonk" call enables the birds to stay in contact with other flock members.

Canada geese feed by dipping from the water's surface, dabbling below the surface and grazing. They forage in wetlands, grasslands and cultivated fields, feeding on grasses, bulbs, grains, berries, seeds of grasses and sedges, and aquatic invertebrates. They eat salt marsh grasses by either grazing on the live grass during the summer and fall or digging up the plant and eating the roots in the late fall and winter.

The geese mature by age three, forming long-term pair bonds. They nest near bodies of water or just up from the tidewater. The female lays five to six eggs, which she incubates for 30 days. In areas where several pairs of geese have been nesting, all of the fledglings are grouped together to form a protective nursery, guarded from predation by several parent birds.

In mid-summer, the geese go through a complete molt, shedding all of their flight feathers. During this time they cannot fly and are vulnerable to predators, and so stay close to the water for quick escape.

Relationship to People

Historically, wild Canada geese were seen in Narragansett Bay only during migration, and hunters would keep tame, wing-clipped geese as live decoy lures. The Migratory Bird Treaty Act of 1918 made it illegal to keep wild birds, and hunters from all over the United States released their decoys, creating large local breeding stocks. As a result, the breeding range is widespread and the species has become so abundant that the geese have become pests.

In areas where many geese congregate near small bodies of water or fields close to swimming beaches or water supplies, their droppings pollute the water, and the birds overgraze the fields, causing soil erosion. To avoid having large flocks gather in small areas for both conservation and public health reasons, a statewide ban of feeding waterfowl was passed in 2003.

FIELD MARKINGS:
Black head and long neck with a distin-guishing white "chin strap" patch from ear to ear. Heavy-bodied with dark wings, pale breast and white under the tail. Males are slightly larger. Size: 25-45 inches long.

HABITAT:
Lakes, ponds, bays, marshes, fields.

SEASONAL APPEARANCE:
Year-round, with additional overwintering flocks.

SENSITIVITY LEVEL:

Common Eider

SCIENTIFIC NAME:
Somateria mollissima

Distinguishing Features and Behaviors

The common eider — the largest duck in the northern hemisphere — is a diving duck with a distinctive sloped forehead, black body and white breast and back. The eastern female appears more red-brown than her western counterpart. Juveniles resemble females but are grayer overall. Eclipse males have dark heads, backs and breasts; first winter males have dark heads with white backs and breasts.

Female common eiders lead their young to water and are often accompanied by nonbreeding hens that participate in chick protection. Broods come together to form "créches" of a few to over 150 ducklings. Attacks by predators may cause several broods to cluster together into a créche. Once formed, a créche tends to stay together throughout the brood-rearing period, although some of the different females attending it may leave.

In winter, common eiders are often seen in rafts of large flocks in the southern part of Narragansett Bay. They eat shellfish, echinoderms, crustaceans and a few fish by diving 33-60 feet underwater.

Common eiders do not nest in Rhode Island yet, but there is evidence the population is expanding southwards naturally and through introductions into Massachusetts. In the past several years, eiders have been seen nesting on several of the Boston Harbor islands, and a nest was recently found on Fishers Island in Long Island Sound.

Relationship to People

Market hunting reduced the common eider population along the East Coast of North America to near extinction by the end of the 19th century. The Migratory Bird Convention in 1916 designated special protection to the eider and significantly reduced the commercial hunting pressure.

FIELD MARKINGS:
Sloping forehead, black body with white breast and back. The wings are white with black primary and secondary feathers; the tail and rump are black. Bill is dull yellow to gray-green. Size: 23 to 27 inches long; wingspan of 35 to 42 inches.

HABITAT:
Open water, ponds, bays.

SEASONAL APPEARANCE:
Late fall, winter, early spring.

SENSITIVITY LEVEL:

Common Goldeneye

COLLOQUIAL NICKNAMES:
Goldeneye, Whistler

SCIENTIFIC NAME:
Bucephala clangula

Distinguishing Features and Behaviors

Common goldeneye are large diving ducks most often seen on the Bay in winter. Males are mostly white with a black back. They have a peaked, glossy green head and a prominent round, white spot before the eye. Females are dusky gray with a dark brown head, white patches on the wings and a white neck collar.

This species may be confused with the Barrow's goldeneye *(Bucephala islandica)*, a rare visitor to Narragansett Bay. The hens of the two species are almost identical except for a more abrupt forehead on the Barrow's. The male Barrow's goldeneye has a crescent-shaped white spot and a glossy purple head.

Common goldeneye are excellent swimmers, using their large and powerful feet to propel them underwater. Their bodies are compact with short wings, thus requiring a long takeoff. In Narragansett Bay, these ducks feed on small crustaceans and mollusks. The common goldeneye uses its diving ability to reach the Bay floor to forage for mussels or small crabs that it crushes with its small, sturdy beak or swallows whole.

This bird acquired the name "whistler" from the loud "whirr" made by its wings in flight. An observer can easily identify flocks of goldeneye arriving to a roost by this distinctive sound.

Relationship to People

Like many diving ducks that spend the winter in Narragansett Bay, common goldeneye rely on the relatively sheltered waters and abundant shellfish. Goldeneye spend the entire winter in the Bay if food is available and hunting pressure is not too extreme. Females and males tend to flock together during their stay, and in spring they fly northward to breeding grounds in the Arctic.

Common goldeneye are hunted in many of the salt marshes and coastal ponds of Narragansett Bay and the Rhode Island coast during winter months.

FIELD MARKINGS:
Males are mostly white with a black back. They have a peaked, glossy green head and a prominent round, white spot before the eye. Females are dusky gray about the body with a dark brown head, white patches on the wings and white neck collar. Size: 20 inches long.

HABITAT:
Open water in bays and salt ponds.

SEASONAL APPEARANCE:
Migratory; late fall, winter, spring.

SENSITIVITY LEVEL:

Common Loon

COLLOQUIAL NICKNAMES:
Loon

SCIENTIFIC NAME:
Gavia immer

Distinguishing Features and Behaviors

Loons are large swimming birds similar in appearance to cormorants. Known for their haunting, laughing call often heard echoing over northern New England lakes and ponds during the summer months, loons are noticeably silent in winter. Unlike the distinct black and white checkered breeding plumage typically seen during the summer months, the loons in Narragansett Bay are in their drab, or eclipse, plumage.

Amazingly adapted for swimming and fishing, loons have been recorded at depths as great as 600 feet. They have long, streamlined bodies with large, webbed feet near the tail. Because their strong legs are placed so far back on the body, loons have great swimming power but are quite clumsy on land.

Loons are equipped with long, daggerlike bills. Their eyes, adapted for underwater vision, make them particularly adept fishing birds. In the daytime, loons fish independently of one another, while at night they congregate, rafting together for protection. They feed on invertebrates and fish, including herring and mackerel.

Loons migrate south along the Atlantic coast from Canada, Maine and New Hampshire, using Narragansett Bay as a wintering spot or staging area. They require clean, healthy, well-oxygenated wetlands and ponds for nesting during the summer months.

Relationship to People

Common loons can be seen on Rhode Island Sound and on Narragansett Bay off Jamestown, Newport and Point Judith during the winter months and occasionally during the spring in their breeding plumage.

They are threatened by the destruction and pollution of their wetland and salt marsh habitat. In 1996, a devastating 828,000-gallon oil spill off the Rhode Island coast was directly responsible for fouling and killing hundreds of loons.

Loons nest on lakes in northern New England that have high levels of mercury due to air pollution. As the top predator in the food chain, the loon has higher concentrations of mercury due to consuming many smaller organisms that have the mercury in their system. The mercury affects the loon's behavior at all stages of its life cycle and reduces its reproduction rates.

Another anthropogenic issue that affects loons is the use of lead sinkers by fisherman. Loons ingest pebbles into their digestive tract to aid them in processing their food. Loons mistakenly swallow the lead sinkers instead of pebbles, causing them to suffer from lead poisoning and, sometimes, death.

FIELD MARKINGS:
Eclipse plumage is mostly dark above, with a white chest and underside. Breeding plumage shows a distinct black head and checkered back with a broken white collar. Size: 28 to 36 inches long.

HABITAT:
Open water in winter months; ponds, streams, wetlands in breeding months.

SEASONAL APPEARANCE:
Winter, spring; rare in summer.

SENSITIVITY LEVEL:

239

Double-crested Cormorant

COLLOQUIAL NICKNAMES:
Cormorant

SCIENTIFIC NAME:
Phalacrocorax auritus

Distinguishing Features and Behaviors

Double-crested cormorants are long-necked black birds often seen sitting on piers and rocks in an upright position with wings in a "spread eagle" posture. The yellow-orange, unfeathered throat pouch of the double-crested cormorant distinguishes it from similar species. Adults in breeding plumage have a crest on either side of the head. Immature birds are dull brown to whitish on the chin, throat and foreneck. In flight, the tail and head appear long and the neck crooked. Cormorants often fly low near the water's surface; a traveling flock may form a line or wedge shape, but they fly silently, unlike noisy flocks of geese.

Cormorants feed on fish and aquatic invertebrates. They dive underwater and swim after fish using their strong paddlelike feet. The end of their bill is hooked to enable them to grasp onto fish.

The outer layer of feathers on cormorants is not water-proof and, when wet, it adds weight to the bird, making it easier to dive. The inner layer stays dry to provide insulation. After diving, cormorants stand on rocks or pilings and spread their wings out to dry the soaked flight feathers. Their eyes are adapted for both aerial and under-water vision. With a soft eye lens and strong eye muscles, they are able to change the lens to a more rounded shape to focus at shorter distances underwater.

Double-crested cormorants nest in colonies on islands in Narragansett Bay. Nests are built in trees or on the ground and are made of sticks, seaweed and debris. A similar species, the greater cormorant *(Phalacrocorax carbo)* is larger with a whitish breast and is seen more commonly along Rhode Island's southern shore.

Relationship to People

Before 1981, there were no nesting birds in Rhode Island, and double-crested cormorants were only sighted migrating to their southern wintering grounds. Since becoming established in the state, double-crested cormorant numbers have been fairly stable, with approximately 2,000 nests per year in Narragansett Bay since the early 1990s.

Cormorants may compete with native wading birds for nesting sites in live trees. When cormorants nest in trees on isolated coastal islands, their dropping (guano) is acidic and destroys the leaves, eventually killing off trees in the nesting colonies. Some small islands in the Bay have been almost completely defoliated by cormorant droppings.

FIELD MARKINGS:
Dark-colored with a long, slender orange bill and blue-green eyes. Feathers have a coppery iridescence when seen up close. Size: 36 inches long, with a wingspan of 52 inches.

HABITAT:
Rocky coast, islands, bays, lakes, rivers.

SEASONAL APPEARANCE:
Year-round, plus increased overwintering population.

SENSITIVITY LEVEL:

Greater Scaup

SCIENTIFIC NAME:
Aythya marila

Distinguishing Features and Behaviors

Greater scaup are small diving ducks, different from dabbling ducks such as the mallard. On the water, male scaup are dark around the head and tail with whitish sides. The similar-looking lesser scaup (Aythya affinis) shows a peaked head and metallic purple color. Female scaup are mostly dark brown with a distinct white mask at the base of a bluish bill. In flight, scaup are identified by a distinct white stripe on the trailing edge of the wing. Because scaup have such short wings compared to the length of their body, they need a long, running start to achieve flight.

Scaup may be observed in large groups, or "rafts," in the open water of upper Narragansett Bay. With a quick upward and forward lunge, scaup dive headfirst into the water. Their large feet enable them to dive to depths of 20 feet and remain underwater for up to one minute. The greater scaup's diet consists of aquatic invertebrates, such as amphipods, which it strains from the water with a specialized bill.

Relationship to People

Greater scaup are common throughout winter in Narragansett Bay. In December 1992, 12,000 scaup were counted on the upper Bay and, currently between 5,000 – 8,000 scaup winter here. They are often seen in the water off Conimicut, Gaspee and Nyatt Points.

In late winter and early spring, many Atlantic flyway migrants such as greater scaup use the Bay as a stopover en route to their northern breeding grounds. When open or unfrozen water is available in the winter — as is quite common for the Bay — scaup will remain here throughout the season.

FIELD MARKINGS:
Rounded, dark head and light-colored sides. Female scaup are mostly dark brown with a distinct white mask at the base of a bluish bill. Size: 16 to 20 inches long.

HABITAT:
Coastal ponds, bays.

SEASONAL APPEARANCE:
Fall, winter, early spring.

SENSITIVITY LEVEL:

Mallard

SCIENTIFIC NAME:
Anas platyrhynchos

Distinguishing Features and Behaviors

Mallard ducks are the most common ducks in Rhode Island. These dabbling ducks use their broad bills in combination with large tongues to sieve through the water and mud, consuming aquatic vegetation, insects and invertebrates. When reaching underwater, their tail end tips up. On land, mallards feed on seeds, shoots, grass, acorns and grain.

Like many other dabbling ducks, mallards are able to spring directly into flight without requiring a takeoff run. This is because they have large wings relative to their body weight.

Mallards pair up for breeding season but choose a new mate each year. Males set up a territory around the breeding female to guard her until a week into incubation. The female builds the nest on the ground, usually in well-concealed vegetation or hollowed logs near the edge of a pond, marsh or water body. Nests are constructed of dry grass, reeds and cattails.

At the end of the breeding season, before migration, mallards molt and lose their flight feathers. For about a month, the birds are flightless. During this vulnerable period, while putting on a layer of fat for migration, they seek out safe or secluded bodies of water for shelter.

Relationship to People

Mallards — uncommon summer residents in Rhode Island in the early 1900s — are now one of the most common nesting waterfowl in the state. The local breeding stock originates from releases and escapees from captive flocks in the early to mid-1900s. Their population continues to expand, benefited by adaptation to human influence.

By filter feeding in water, ducks consume pollutants along with their intended foods. Trace metals and other elements are retained in their feathers and body tissues. A major threat to dabbling ducks was the use of the lead shot, which was banned in 1991 by the U.S. Fish and Wildlife. Feeding ducks consumed the small, lead pellets, which entered the blood stream after digestion and often caused their death. After the ban, researchers recorded a reduction in the mortality rate of mallards due to lead ingestion.

FIELD MARKINGS:
Males have a metallic green head with a white collar, rusty breast, green under-side, black and white tail, yellow bill and orange feet. Females are mottled, tawny brown with a whitish tail and black-orange bill. Both sexes have a violet-blue bar bordered with white lines on the wings. Size: 23 inches long.

HABITAT:
Marshes, wooded swamps, grain fields, ponds, rivers, lakes, bays.

SEASONAL APPEARANCE:
Year-round, plus additional overwintering population.

SENSITIVITY LEVEL:

245

Mute Swan

SCIENTIFIC NAME:
Cygnus olor

Distinguishing Features and Behaviors

The mute swan is a large, white bird common in both the salt and fresh waters of Rhode Island. They are not mute as the name implies; rather, they make hissing sounds while guarding territory. With its bill pointed down when swimming, the swan's long neck forms a distinctive "S" shape, making it easy to identify from a distance.

Mute swans fly with their necks extended, and the stiff feathers of their wings create a swooshing noise that can be heard up to a half mile away. Due to their weight, they require a long runway to become airborne. Mute swans feed on pond weed, microalgae and aquatic invertebrates. They graze on salt marsh grasses and in the late fall and winter will uproot the plants and feed on the roots. Swans can consume eight pounds or more of vegetation per day. They feed like dabbling ducks, tipping forward and extending their long necks into the water. They have small toothless edges on the inside of their bills to grasp vegetation.

Mute swans pair up for life when they are three to four years old. Both the male and female build nests in marshes, brackish ponds or shallow water out of cattails, cordgrasses and *Phragmites*. Males are territorial, often going to great lengths to defend their nests aggressively. Swans often swim with their heads back and wings arched. This behavior, called "busking," is an aggressive display to defend their territory. During the first year of life, the gray, immature swans, or cygnets, can be seen following the adults in close proximity.

Relationship to People

Mute swans are not native to North America. They were imported from Europe to the New York City area in the late 1800s for their beauty. The introduced population increased in the early 1900s when escapees began to breed in the wild. The first swans in Rhode Island were seen in Quonochontaug Pond, Charlestown, in 1938.

Swans have large breeding territories and aggressively drive off native birds from their nesting habitat, often altering the diversity of birds in an area and affecting nesting success. In areas where swans gather in large numbers, they can degrade water quality through intensive feeding and high-nutrient waste. Since the first nesting in 1948, the number of mute swans has been on the rise despite efforts to control populations including "addling," or shaking the eggs.

FIELD MARKINGS:
All white feathers on the body, red-orange bill with a black knob at the base, black legs and feet. Female is slightly smaller than male and has a smaller black knob on the bill. Size: 60 inches long, with a wingspan of 60-plus inches.

HABITAT:
Fresh and saltwater ponds, coastal marshes, coves on the Bay in winter.

SEASONAL APPEARANCE:
Year-round.

SENSITIVITY LEVEL:

Red-breasted Merganser

SCIENTIFIC NAME:
Mergus serrator

Distinguishing Features and Behaviors

Mergansers are large diving ducks that feed almost exclusively on fish. Their specially adapted, spikelike, serrated-edged bills are particularly useful in catching and devouring fish. These attractive birds — often seen fishing in pairs or groups — work together to drive fish into shallower water, making the fish easier to catch. When on the water, mergansers ride low and take frequent dives.

The similar-looking common merganser *(Mergus merganser)* female differs from the red-breasted merganser in that the white throat is well defined against the colored head. Common mergansers are more common in freshwater tributaries.

Red-breasted mergansers do breed in Narragansett Bay habitats; however, most birds seen on the Bay are only winter visitors. They build their nests in a scraped-out bowl on the ground, usually within 25 feet of the water and well-concealed by tall grass, tree roots or woody vegetation.

A nesting pair of red-breasted mergansers normally produces eight to ten eggs. The female will incubate for approximately 30 days before hatching downy and mobile young.

Relationship to People

Rhode Island boasts a large winter population of red-breasted mergansers, numbering in the hundreds, and occupying many inlets and coastal ponds of Narragansett Bay. These birds withstand harsh New England conditions by occupying sheltered embayments and coves. They hunt for fish such as sculpin, herring and mummichogs.

Red-breasted mergansers are hunted heavily during their migration through Rhode Island. In 1993, 1,066 birds were taken in a five-day hunting season.

FIELD MARKINGS:
Males have a glossy green head with a distinct double crest and broad white collar above a chestnut breast. Females are mostly gray with white throat and chin and crested rusty head. Size: 20 to 26 inches long.

HABITAT:
Open water of bays, coastal ponds.

SEASONAL APPEARANCE:
Year-round; more abundant in winter.

SENSITIVITY LEVEL:

Black-crowned Night Heron

Scientific Name:
Nycticorax nycticorax

Distinguishing Features and Behaviors

The black-crowned night heron is a sturdy-looking heron with a short bill, neck and legs. When perched, it has a distinguishing hunched posture with its head drawn down to meet its shoulders. In flight, this heron curls its neck back so that its head and back form a straight line and the feet barely extend beyond the tail. With deep strokes of its arched wings, the black-crowned night heron can fly at 20 to 35 miles per hour. It also swims well and lands on the surface of the water like most other herons.

Black-crowned night herons are nocturnal, feeding mainly between dusk and dawn, thus the name "night herons." Their Latin name *Nycticorax* means "night raven." These birds have an adaptable diet, eating whatever is plentiful. Their diet consists of fish, insects, amphibians, eggs, the young of other birds (including terns, herons and ibises) and small mammals. Like other herons, they are experts at still-fishing: standing motionless until a fish comes within range before striking out to catch it.

Because night herons roost in trees during the daytime, they are fairly easy to spot, not only by bird watchers, but also by day herons, which try to attack the night herons and drive them off.

Relationship to People

Black-crowned night herons were more common at the beginning of the 20th century with colonies on the large islands of Narragansett Bay and on the mainland. The Bay population declined due to disturbance of their rookeries, hunting and the use of DDT and other harmful pesticides in the 1950s and 1960s.

Today, black-crowned night herons nest colonially in trees and shrubs on isolated islands in the Bay. Peak nesting numbers occurred in 1992 when 496 black-crowned night heron nests were detected, with the largest colony site located on Hope Island. Numbers have been declining since that time.

Currently, the North American Waterbird Conservation Plan lists black-crowned night herons as a species of moderate conservation concern. The State of Rhode Island recognized these herons as a species of conservation concern due to the limited extent of nesting habitats.

FIELD MARKINGS:
A stocky heron with a black crown contrasting a mostly gray body. A distinct black bill, yellow legs and brilliant red eyes. Size: 25 inches long, with a wingspan of 44 inches.

HABITAT:
Marshes, wooded wetlands, marine islands.

SEASONAL APPEARANCE:
Spring, summer, fall.

SENSITIVITY LEVEL:

Glossy Ibis

SCIENTIFIC NAME:
Plegadis falcinellus

Distinguishing Features and Behaviors

Glossy ibises are wading birds common to the marshes of Narragansett Bay. They appear dark at a distance; however, adults are chestnut-colored with an iridescent purple gloss on the head, neck and underside. During the nonbreeding season, the facial skin is gray, while in the breeding season it becomes a cobalt blue, with the leg joints appearing red on green-gray legs. Immature birds are dark green with brownish heads, and their necks are covered with streaks. Unlike heron and other wading birds, ibises fly with their necks extended.

Ibises wade through the shallows of fresh and saltwater marshes, probing the mud for crabs, crayfish, worms and other aquatic invertebrates. They feed commensally with other wading birds, such as the snowy egret. Commensal feeding is when one bird helps another find food without wasting any extra energy. Since the ibis forages by feel, it stirs up food in the murky waters for the egret, who forages by sight. The ibis also relies heavily on agricultural lands for feeding, especially when feeding young during the late spring months.

Ibises nest in small mixed colonies with other wading birds, such as egrets and night herons, on islands in the Bay. Nests are built of sticks and twigs in trees, dense thickets and shrubs up to ten feet from the ground, or sometimes even *on* the ground. A lining of dry material is maintained within the nest until the young leave.

The female lays three to four eggs, and both parents incubate for 21 days; the female incubates at night and the male during the day. After about seven weeks, the young leave the nest and forage with the parents.

Relationship to People

Glossy ibises are originally from Africa and immigrated to the South American continent in the 19th century. They were first observed nesting in the United States in 1880 and are now found as far north as Maine and the Great Lakes. First spotted in Rhode Island in the 1930s, the first recorded nesting in this state was in 1972. Their population reached a high of 521 nesting pairs in 1991, but has fluctuated from 68 to 301 nests annually since then.

This bird is listed as a species of conservation concern in the State of Rhode Island. The greatest threat to the glossy ibis is habitat destruction of their feeding and nesting sites.

FIELD MARKINGS:
Dark, copper-colored wading bird with a long, down-curved bill and long, gray-brown legs. Size: 23 inches long, with a wingspan of 36 inches.

HABITAT:
Brackish and saltwater marshes, estuaries, coastal islands, fields.

SEASONAL APPEARANCE:
Spring, summer, early fall.

SENSITIVITY LEVEL:

Great Blue Heron

COLLOQUIAL NICKNAMES:
Blue Heron, Great Blue

SCIENTIFIC NAME:
Ardea herodias

Distinguishing Features and Behaviors

The largest heron in North America, great blue herons are commonly seen along the edges of coastal waters and rivers. In Florida some adults are all white, a regional color morph, and were once thought to be a separate species. In flight, great blues are recognized by their size and silhouette: long necks tucked in and slow, sweeping wing beats; they hover to a shallow-water landing. These herons are lighter than they appear, with an average adult weight of five to eight pounds.

Great blue herons forage for fish and amphibians by sweeping the water with open bills until contact with the prey triggers a reflex to snap the bill closed. They will also stand motionless, waiting to stab at swimming prey. Their diet consists mainly of fish, but they also eat human food scraps, crabs, crayfish, nestlings and small mammals.

These herons nest together in colonies, sometimes with other species of wading birds. They build their nests up to 130 feet from the ground in dead trees or in shrubs found in flooded, forested areas. Great blues forage in coastal habitats year-round, except for May and June when they are rarely detected in the Bay.

A similar heron species seen increasingly along the southern coast of the Bay is the little blue heron *(Egretta caerulea)*. Smaller than the great blue – at 24 inches long and with a 42-inch wingspan – little blues have bluish-gray bodies and purplish heads and necks. The main difference between the herons is in how they feed. Larger herons appear more graceful, while the little blue heron awkwardly stabs at prey. They often feed in the company of other wading birds, which increases their chances of catching prey disturbed by the other birds.

Relationship to People

Historically, there may have been a greater nesting population of great blue herons in Narragansett Bay in the early 1900s and late 1800s when beavers were more prevalent. When streams are dammed by beavers, the low-lying areas are flooded. This kills trees and provides nesting habitat for the herons.

Given the limited nesting habitat available, the State of Rhode Island lists this bird as a species of conservation concern. Population numbers have been increasing, but there has also been increased habitat destruction of many nesting areas in the region.

FIELD MARKINGS:
A large grayish-blue wading bird with a long, white neck with black streaks. Size: 46 inches long, with a wingspan of 72 inches.

HABITAT:
Marshes, wetlands, shores, tidal flats.

SEASONAL APPEARANCE:
Summer, fall; some birds overwinter.

SENSITIVITY LEVEL:

Great Egret

COLLOQUIAL NICKNAMES:
Common Egret,
American Egret

SCIENTIFIC NAME:
Ardea alba

Distinguishing Features and Behaviors

Great egrets often flock with other wading birds and are the largest white-colored herons found in this region. In the breeding season, great egrets have long plumes that fall gracefully from the back of the bird, extending beyond the tail. These feathers are lost by the summer months. Male and female birds have similar plumage. Immature birds are all white but lack the long plumes, and their legs and bills are duller in appearance than those of adults. Great egrets are identified by their broad white wings and slow, sweeping flight. Their long necks are tucked in as they fly with their legs trailing behind.

Egrets prefer to forage in the open areas of tidal flats and salt marshes at low tide so they can stalk their prey in a slow, methodical way. Their diet includes fish, insects, amphibians and small birds.

Great egrets can be seen before dusk as they gather to roost for the night. They nest in colonies, sometimes with other species of wading birds, on islands in the Bay. Nests — consisting of sticks and twigs, sometimes lined with leaves or other finer vegetation — are constructed in trees and shrubs from ten to 30 feet above the ground.

Relationship to People

In the early 1900s, the great egret population was significantly reduced due to the millinery trade. The long, white plumes of great and snowy egrets were quite valuable, as it was fashionable to adorn ladies hats with the large feathers. In 1903, plume hunters were getting $32 an ounce for the ornate plumes. After the 1918 Migratory Bird Treaty Act, which protected nongame birds from being hunted and their eggs and nests from being collected, the egret population recovered and expanded its range. Now egrets are threatened by pesticides and the loss of their nesting and feeding habitats by human development.

Great egrets currently nest on four islands in Narragansett Bay. Annual nest numbers peaked with 233 nests in 2003, but the availability of habitat for this species remain a cause for concern. The great egret is listed as a species of conservation concern in the State of Rhode Island.

FIELD MARKINGS:
A large, all-white heron with a long neck, black legs and feet and a long yellow bill. Size: 39 inches long, with a wingspan of 51 inches.

HABITAT:
Marshes, mud flats, ponds, coves.

SEASONAL APPEARANCE:
Spring, summer, early fall.

SENSITIVITY LEVEL:

Snowy Egret

SCIENTIFIC NAME:
Egretta thula

Distinguishing Features and Behaviors

Snowy egrets are medium-sized "snow white" herons. Both sexes have similar plumage. Immature birds are white with yellow feet and yellow-green legs. Due to their size, snowy egrets fly faster than great egrets and are much more maneuverable in the air. During the breeding season, adults are adorned with long, white, upturned plumes on the back, neck and head; at this time, the area between the eye and the base of the bill, (the lores) turns bright red, and their feet turn golden.

During the day, snowy egrets can be found foraging in marshes, ponds and shallow coastal waters. Unlike other wading birds that stand and wait for their prey, snowy egrets hunt actively. Using their yellow feet as lures to attract fish, they move hurriedly about in the water, stirring up the bottom and catching their prey with repeated stabs. Snowy egrets take advantage of feeding opportunities presented by other wading birds that herd fish into shallow waters. They hunt by hovering above the water and dropping down on their prey or by running along the shallows. Their diet consists of fish, insects, aquatic invertebrates and small vertebrates. These birds gather with other wading birds to roost for the night.

Snowy egrets nest in mixed colonies with other wading birds on the islands of Narragansett Bay. They build their nests of sticks in trees or shrubs between five and ten feet off the ground.

Relationship to People

In the late 1800s and early 1900s, snowy egrets were slaughtered by hunters in great numbers for their small, curved plumes. They were hunted more than great egrets because their numbers were more abundant and their feathers considered more valuable.

Protection by the Migratory Bird Treaty Act has allowed this species to recover and expand its range to the point where it is now one of the more common wading birds in Rhode Island. Snowy egrets began nesting on Little Gould Island in 1965. Since that time, colonies have been established on three to four islands within Narragansett Bay. Annual nest numbers peaked at 223 in 1991, but have since declined, not rising above 100 nests since 2000. The snowy egret is listed as a species of conservation concern in the State of Rhode Island and is recognized as a high risk species by The North American Waterbird Conservation Plan because of large regional declines in recent years.

FIELD MARKINGS:
Mostly "snowy" white. Black bill, black legs and bright yellow feet. Size: 24 inches long, with a wingspan of 41 inches.

HABITAT:
Marshes, ponds, shallow coastline, tidal flats, sometimes dry fields.

SEASONAL APPEARANCE:
Spring, summer, early fall.

SENSITIVITY LEVEL:

Greater Yellowlegs

SCIENTIFIC NAME:
Tringa melanoleuca

Distinguishing Features and Behaviors

Greater yellowlegs are medium-sized shorebirds with long yellow legs and long necks and bills. Although similar in appearance to another shorebird, the lesser yellowlegs *(Tringa flavipes)*, the greater is larger and has a longer bill.

Yellowlegs forage for food in tidal areas. They wade into the water, darting around the shallows and stabbing the water for prey. Unlike other shorebirds with long bills, greater yellowlegs do not probe the sand for invertebrates; rather they use a swaying motion to strain the water with their bill.

Wading birds like egrets and ibises stir up food in the shallows, and the yellowlegs take advantage by excitedly darting around and snatching up small fish. They are often described as "frantic feeders." Small invertebrates and berries are also included in their diet.

Most species of shorebirds seen in Rhode Island are migrants and overwintering birds that do not breed here. Shorebirds such as the greater and lesser yellowlegs nest in northern Canada and Alaska. They pass through Rhode Island on their way to wintering grounds in the southern United States and South America. The fringe marshes of Narragansett Bay serve as preferred habitat during migration.

Greater yellowlegs are often seen standing on one leg with the other leg tucked up underneath their feathers. Yellowlegs will shift the foot they are standing on as the foot gets cold. This behavior helps to prevent heat loss through the skin surface on areas that are not feathered.

Relationship to People

Before 1900, some shorebirds were hunted in great numbers. Although hunting of nongame species is against the law, there are different threats to their population. Shoreline development is threatening these birds due to habitat loss and increased pollution of feeding areas. Domestic animals like dogs and cats cause increased stress to the greater yellowlegs as well as direct predation on its population.

Contamination of the shoreline and coastal ponds by oil and other toxic pollutants kills the aquatic invertebrates that sustain the greater yellowlegs during their stopovers in the Bay and along the coastal shore.

FIELD MARKINGS:
Dark gray body and long yellow legs. Long upturned bill is grayish at the base and dark at the tip. In breeding plumage, the upper part of the body is mottled with black, gray-brown and white and has streaks along the throat and breast. Size: 12 to 15 inches long, with a wingspan of 23 inches.

HABITAT:
Tidal flats, marshes, shallow wetland areas, shores.

SEASONAL APPEARANCE:
Spring, late summer, fall.

SENSITIVITY LEVEL:

Killdeer

SCIENTIFIC NAME:
Charadrius vociferus

Distinguishing Features and Behaviors

The killdeer is a large banded plover with brown upperparts, white underparts and two distinct black bands cross the upper breast. In flight, the tail and rump show rust brown, and white stripes are visible on its wings. The male and female are similar. The juvenile has a single breast band.

Since conservation measures were enacted, the killdeer has become the most common nesting shorebird in Rhode Island. They are ground-nesting birds famous for "hiding" their nests right out in the open. They use no nesting materials and rely on distraction displays to protect their offspring.

The killdeer produces four dark-spotted, buff eggs that are laid in a shallow ground depression lined with grass, often in gravelly areas. Killdeer exhibit a clever "broken wing display" in which they appear to be struggling with a broken wing all the while leading the predator away from their babies. Once their young are out of danger they "recover" and fly off.

Their diet consists of various aquatic and terrestrial invertebrates, mostly insects. They also eat berries and forage on the ground.

Killdeer emit a loud "kill-dee" or "kill-deear" or "kill-deeah-dee-dee" cry. They also make a long, trilled "trrrrr" sound during display or when their young are threatened.

Relationship to People

The killdeer will nest just about anywhere — parking areas, sidewalks, abandoned lots — so they are relatively unperturbed by humans.

FIELD MARKINGS:
Large banded plover with brown upperparts, white underparts and two distinct black bands cross the upper breast. Size: 9 to 10.5 inches; wingspan 19 to 21 inches.

HABITAT:
Fields, river banks, shores, uplands.

SEASONAL APPEARANCE:
Year-round.

SENSITIVITY LEVEL:

Peep Sandpipers

Least Sandpiper

Distinguishing Features and Behaviors

Three species of small sandpipers — unofficially categorized as peep sandpipers — are so similar in appearance, they are a challenge to identify.

The easiest way to identify each bird is to watch for characteristic behaviors, observe them in their preferred feeding habitats and note subtle field marks.

Sanderlings feed at the shore's edge, running in and out of the water, probing the sand with every outgoing wave. When in flight, the wing of the sanderling has a broad white stripe. Least sandpipers have yellowish legs and slightly decurved or down-curved bills, and they feed along salt marsh creeks. Semipalmated sandpipers have black legs and straight, blunt-tipped bills and feed closer to the high tide line, searching through washed-up seaweed in a hunched-over posture.

Sandpipers feed on insects, aquatic invertebrates, snails and small crustaceans. Each species — with a slightly different length and shape of bill — adapts to its particular way of probing; some can be found feeding together in flocks, consuming different food sources.

Most sandpipers are in Rhode Island only as migrants, with few staying through the winter. In spring they nest in the Arctic tundra where they take advantage of the abundant supply of insects. In mid- to late summer, sandpipers leave their nesting grounds and fly south to spend winter along the coasts of Central and South America.

Sandpipers eat as much food as they can to store up necessary fat reserves for their long migratory journey. Every time a flock of birds is disturbed or stressed, energy is wasted and the birds may not be strong enough to complete their trip.

Shorebird flocking behavior can be quite spectacular. If a flock is disturbed, all the birds fly up in close formation, moving as a group in synchrony, then landing together. This behavior helps protect individual birds from aerial predators such as hawks and falcons.

Relationship to People

Even though these shorebirds are present in great numbers, their survival is dependent on the conservation of key migratory stopover sites. With coastal and shoreline development encroaching on these staging areas, people and their domestic animals threaten shorebird survival.

FIELD MARKINGS:
Plump with a brownish or gray-brown mottled head, wings and back and a white or light underside. Thin, sticklike legs move fast as they run along the shore. All three species have dark, tubular bills that vary only slightly in length and shape. Size: 6 to 8 inches long.

HABITAT:
Shoreline, tidal flats, marsh areas, river sand bars.

SEASONAL APPEARANCE:
Spring, late summer, fall; few overwinter.

SENSITIVITY LEVEL:

Piping Plover

SCIENTIFIC NAME:
Charadrius melodus

Distinguishing Features and Behaviors

Piping plovers are small and plump, sand-colored shorebirds. They are difficult to spot on the beach due to coloration that blends in with the sand. They have adapted to concealing themselves by moving quickly across the sand, then stopping abruptly and crouching to become virtually invisible.

Plovers feed on invertebrates found along the shore-line, such as small crustaceans, mollusks, worms and invertebrate eggs. A similar-looking local bird is the semipalmated plover *(Charadrius semipalmatus)*, which can be distinguished easily by its overall brown-ish color and distinct black mask.

Piping plover pairs form long-term bonds and return to the same nesting site each year. Both birds help build the nest, which is a simple depression or scrape in the sand, lined with pebbles or shells for camouflage. Nests are located above the high tide line in sandy areas with sparse vegetation. Piping plovers usually have one brood per year, with four eggs. After an incubation of 25 to 30 days, young birds hatch. They are then mobile and able to find their own food, following the parents closely for safety and crouching in the sand when alarmed by a possible predator. Young birds are able to fly 20 to 32 days after hatching.

This plover's call is a series of melodious whistles described as "peep-lo."

Relationship to People

The piping plover suffered population losses before protection measures were put into place in 1918. The population recovered only to decline again due to loss of suitable breeding habitat. They were included on the Blue List from 1972 to 1982, and the eastern North American population was listed as a federally threatened species in 1986. At that time, it was estimated that there were only 4,500 birds left in existence, with none breeding in Narragansett Bay.

Through protection, piping plovers are now breeding in the Bay. However, they are still threatened by habitat loss from shoreline development, oil spills and the danger of their nesting sites and young being destroyed by recreational vehicles and unwary pedestrians.

FIELD MARKINGS:
White underside, yellow-orange legs and a short orange bill with a black tip. Bill and legs turn dark in the winter. A narrow band around the neck varies in shade between the seasons. Across the forehead is a bar that is black in the breeding season and fades to a sandy gray during winter. Size: 7 inches long.

HABITAT:
Sandy beaches, dunes, tidal flats.

SEASONAL APPEARANCE:
Spring, summer.

SENSITIVITY LEVEL:

Spotted Sandpiper

SCIENTIFIC NAME:
Actitis macularia

Distinguishing Features and Behaviors

The spotted sandpiper is a small shorebird displaying olive-brown upperparts and white underparts with bold black spots. Its eyestripe is white and the wings have white stripes visible in flight. Its tail is barred. The female spotted sandpiper has larger belly spots while the winter adult is duller in color and lacks spots.

Common along rivers, the spotted sandpiper also can be found along the upper sections of Narragansett

Bay. It is identified easily by its stiff wing beats — each time the spotted sandpiper flaps its wings, they appear to stop in mid-flight. This shorebird also can be found nesting in some areas along the Bay, including in the salt marsh and coastal buffer at the Save The Bay Center in Providence.

The female defends the birds' territory and mates with several males, a practice known as polyandry. The spotted sandpiper produces four brown-spotted, buff eggs, which are laid in a ground depression lined with grass or moss. Incubation ranges from 20 to 24 days and is carried out by the male. The eggs and hatchlings are highly susceptible to predation.

This shorebird eats insects, fish, worms, mollusks, crustaceans and spiders. It forages on the ground or in shallow water, often dipping its food in the water before eating it. Adults and juveniles continuously teeter on their stiltlike legs; this behavior is thought to stir up the small insects it prefers to feed upon.

The spotted sandpiper call is a clear "peet-weet" and also a soft trill.

Relationship to People

Since spotted sandpipers prefer to nest in open fields, their nesting population was likely larger at the turn of the century when most of the Narragansett Bay watershed land use was agricultural.

FIELD MARKINGS:
Olive-brown upperparts and white underparts with bold black spots. Female is similar but has larger belly spots. Size: 7.5 inches long; wingspan of 13-14 inches.

HABITAT:
Salt marsh, mudflats, coastal uplands, fresh water shores.

SEASONAL APPEARANCE:
Early spring, summer, early fall.

SENSITIVITY LEVEL:

Willet

SCIENTIFIC NAME:
Catoptrophorus semipalmatus

Distinguishing Features and Behaviors

The willet is a large sandpiper with mottled gray-brown upperparts, a white rump and lightly streaked and barred white underparts. Broad white stripes on black wings are visible in flight. Its tail is white with a dark brown tip; the legs are blue-gray. Male and female markings are similar. The winter adult is plain gray-brown above and white below.

The willet — one of the few shorebirds that nest in Narragansett Bay — is commonly found in salt marshes. It produces four brown-spotted, olive-brown eggs laid in a nest lined with weeds or bits of shell and built into a depression on open ground or in a grass clump. The male incubates the eggs and will attend to the brood after the female abandons them two to three weeks after they hatch.

This shorebird feeds on mollusks, crustaceans, insects and small fish, and forages by picking food from shallow waters and probing mud with the tip of its bill.

The willet's loud call makes its appearance known long before it is observed. It is a loud, ringing "pill-will-willet" and a quieter "kuk-kuk-kuk-kuk-kuk."

Relationship to People

A fairly recent nester in Narragansett Bay, the willet has expanded its nesting range north in this century. Like other shorebirds, its population is beginning to rebound and expand northward since the end of the millinery trade (see the least tern). The willet is state listed as a species of concern.

FIELD MARKINGS:
Mottled gray-brown upperparts, white rump and lightly streaked and barred white underparts. Broad white stripes on black wings are visible in flight. Tail is white with dark brown tip. Legs are blue-gray. Size: 13 to 16 inches long; wingspan 24 to 31 inches.

HABITAT:
Salt marshes, intertidal zone, marshy lake and pond margins.

SEASONAL APPEARANCE:
Spring, summer, early fall.

SENSITIVITY LEVEL:

Great Black-backed Gull

COLLOQUIAL NICKNAMES:
Great Black-back Gull,
Seagull

SCIENTIFIC NAME:
Larus marinus

Distinguishing Features and Behaviors

The great black-backed gull is the largest species of gull in Narragansett Bay, distinguishable from other gulls by their large size, white head and characteristic black back.

To reach the full, black adult plumage takes four years, with young birds going through stages of different plumage with every molt. Juveniles are identified by their smaller body and bill size – their head and body are mottled with dark gray and the wings are spotted somewhat darker. Compared to a young herring gull, their heads and bodies appear lighter and their wings more spotted.

When gulls of different species gather together to feed, the great black-backed will drive away or steal the other birds' food. They feed on eggs, small birds, squid, carrion, shellfish, fish, berries, grain, scraps from fishing boats and garbage.

Great black-backed gulls nest in colonies with other species, such as herring gulls and double-crested cormorants. They colonize small islands in the Bay and offshore, setting up territories around their nests up to a six-foot diameter. Nests are built on the ground from mounds of seaweed, grasses, vegetation and available debris. The male and female court, forming a pair bond that can last more than one breeding season. The female lays two or three eggs, and both birds take turns incubating for 26 days and feeding the fledglings for another 42 to 49 days.

Great black-backed gulls overwinter along the Atlantic coast from Newfoundland to North Carolina, with large numbers on Cape Cod. Great numbers of young birds can be found farther out to sea where they follow fishing vessels — unlike herring gulls that stay closer to the coastline.

Relationship to People

Because great black-backed gulls are good scavengers and have adapted to living near people, their population is increasing. In Rhode Island, the nesting population reached its peak with 1,920 nests in 1992. Since then, nest numbers seem to be declining, however the regional increase in these gulls may pose a serious threat to other coastal sea birds. The great black-backed gull competes for nesting sites and food and is notorious for robbing nests of eggs and nestlings. They are responsible for reducing tern and eider duck populations by stealing fish from the parent birds as they feed their own nestlings.

FIELD MARKINGS:
Adults have an all-white head and body with distinct black back and wings; large yellow bill, with a bright orange-red spot near the tip of the lower bill. Size: 28 to 31 inches long, with a wingspan of 65 inches.

HABITAT:
Ocean coasts, bays, beaches, piers, landfills.

SEASONAL APPEARANCE:
Year-round, plus overwintering population.

SENSITIVITY LEVEL:

Herring Gull

COLLOQUIAL NICKNAMES:
Seagull

SCIENTIFIC NAME:
Larus argentatus

Distinguishing Features and Behaviors

Herring gulls are the most common gulls on the Atlantic coast. An adult is easily recognized by its overall white body with pale gray wings and back. Gulls can appear so similar, they are commonly lumped into one category called "seagulls." A "seagull" is not a specific species of bird, but rather a common term incorrectly used for all gull species.

Herring gulls do not acquire their distinctive adult plumage until they are four years old, which makes juveniles difficult to identify. In the first three years, their plumage varies from an overall brown with darker, brownish wings to a lighter gray color, gradually changing from season to season. The color of the bill, eyes and legs also changes with the age of the bird, from dark to yellow by the time they reach adulthood.

Most gull species feed on fish and shellfish; however, herring gulls have become so adapted to human society they will eat just about anything. They are amazing scavengers, eating garbage, carcasses, food scraps, shellfish, berries, eggs and other bird species. Herring gulls are pirates and will steal food from other birds whenever the opportunity arises.

Herring gulls nest in colonies on the islands of Narragansett Bay and off the coast, usually in areas inaccessible to ground predators. Colonies are formed on the ground near vegetation, on rocky ledges or sometimes in grassy dunes. Pairs set up a territory around their nests and return to the site every year.

Relationship to People

In 1990, there were over 6,000 herring gull nests detected in Narragansett Bay. More recently, annual nest numbers have ranged from 2,811 in 1998 to 2,286 in 2007. This population increase is likely due to the increase in landfills and the bird's ability to scavenge. Landfill closures and better trash management in Rhode Island are probable causes of the reduced Bay population.

Before 1918, the herring gull population may have been controlled by humans hunting their eggs. Increased populations are a threat to other sea birds, such as terns and smaller gulls. Herring gulls drive birds away from their nesting sites, steal food and prey on chicks and even small adults. Managing and covering landfills so they are not a ready food source is one measure that helps control rising gull populations.

FIELD MARKINGS:
White body with pale gray wings and back; black wing tips with white spots; pink legs and feet. Bright yellow bill with an orange spot. Size: 23 to 26 inches long, with a wingspan of 58 inches.

HABITAT:
Ocean coasts, bays, beaches, piers, farmland, landfills.

SEASONAL APPEARANCE:
Year-round, plus overwintering population.

SENSITIVITY LEVEL:

Least Tern

SCIENTIFIC NAME:
Sterna antillarum

Distinguishing Features and Behaviors

The smallest North American tern, least turns have a slender body, narrow pointed wings and a forked tail. Their size, rapid wingbeats and graceful, fluttering flight maneuvers distinguish them from other terns. Young birds have dark bills and legs and lack the adult's distinct black cap.

Least terns feed mostly on small fish, which they hunt by hovering over the water, sometimes for prolonged periods. Upon sighting a fish, they dive headfirst into the water, catching the prey in their bills. They also feed on insects, crustaceans and other aquatic invertebrates.

Least terns nest singly or in colonies, scraping a nest into the sand or pebbles. They arrive in early spring to nest on sandy beaches and island habitats. Pairs do not roost in the breeding area during the night until it is time to incubate the eggs. This decreases the chances of the nest being found by a nocturnal predator.

The size of this bird does not stop it from attempting intimidation tactics to drive away predators. Like the common tern *(Sterna hirundo)*, they will swoop down on humans or other intruders and defecate on them.

Relationship to People

Least terns suffered great population losses during the millinery trade in the late 1800s and early 1900s when the entire bird was stuffed and mounted on ladies' hats. Plume hunters nearly wiped out the species until protection laws were passed. The population recovered from the millinery trade, but has been in steady decline for the last ten years.

There are currently 14 known nesting sites in Rhode Island. In 2006, approximately 400 least terns nested in the state. With the increase in both shoreline development and gull populations, there has generally been a loss of nesting habitat. Human and animal disturbance can stress the terns to the point that they may abandon their nests altogether.

The least tern is listed by Rhode Island as a threatened species, and portions of the population are on the Federal Endangered Species List. Efforts are underway to provide protection of the birds and their nesting sites. Fencing off and posting known nesting areas is one current measure of protection for this species.

FIELD MARKINGS:
Light gray on the back and wings; white underside and black crown extending down the nape of the neck. White forehead, a slender, dark-tipped yellow bill and yellowish legs. In flight, they show white under the wing and a black line on the outer edge of its tip. Size: 9 inches long, with a wingspan of 8 to 11 inches.

HABITAT:
Sandy ocean beaches, bays, large rivers.

SEASONAL APPEARANCE:
Spring, summer, early fall.

SENSITIVITY LEVEL:

Ring-billed Gull

COLLOQUIAL NICKNAMES:
Seagull

SCIENTIFIC NAME:
Larus delawarensis

Distinguishing Features and Behaviors

The ring-billed gull is one of the most common gulls found in the winter along Rhode Island beaches. Adult birds resemble the common herring gull, but can be distinguished easily by their smaller size and rounded, dovelike heads. During winter, adults have light brownish streaks on their heads and the back of their necks.

It takes three years for ring-billed gulls to acquire their adult plumage. Before then, they are slightly more difficult to recognize in the sub-adult plumage. Young ring-billed gulls are paler than young herring gulls, and their heads and bodies have light gray spots. Juveniles also have dark bills and dark eyes that change as they mature.

Like herring gulls, ring-billed gulls will eat just about anything from fish, bird eggs, insects, mollusks and worms to berries, small rodents, scraps from fishing boats and garbage. They are pirates as well as scavengers and will steal food from other birds.

To reach the inner part of clams, gulls fly over rocks and drop the mollusks on hard surfaces, such as roads or jetties, to break them open. Sometimes this behavior is repeated several times before they are successful.

Relationship to People

The breeding range of the ring-billed gull increases and expands each year. Overwintering populations in Rhode Island benefit because they have adapted to scavenging at area landfills and shellfishing areas. Abundant, easy food sources such as these have helped ring-billed gulls survive through winter, leading to an increased population in their breeding areas.

FIELD MARKINGS:
White body, gray wings and back. Bill is yellow with a black ring, pale yellow eyes and bright yellowish or yellow-green legs. The tips of the wings are black and have two white spots, easily seen when the bird is standing. Size: 18 to 20 inches long, with a wingspan of 48 inches.

HABITAT:
Coasts, bays, beaches, piers, landfills.

SEASONAL APPEARANCE:
Late summer, fall and winter.

SENSITIVITY LEVEL:

Northern Harrier

COLLOQUIAL NICKNAMES:
Marsh Hawk

SCIENTIFIC NAME:
Circus cyaneus

Distinguishing Features and Behaviors

The northern harrier, formerly known as the marsh hawk, is a familiar sight along the coastal fields and salt marshes of Narragansett Bay.

The harrier is a large hawk with gray upperparts, a distinct white rump and white underparts with a spotted breast. Its eyes are yellow, and its wings are long — gray above and white below — with black tips. A large white band along its tail is a clear field identification guide. The female is larger than the male and brown overall, with a white rump and under-wing stripes.

The female feeds and broods the young and aggressively excludes males from feeding areas. Although you can see northern harriers hunting as far north as Fields Point in Providence, Block Island hosts the raptor's only known breeding grounds in Rhode Island.

Harriers produce three to five pale blue eggs which are laid in a ground nest made of sticks and lined with grass, usually built on a raised mound of dirt or in a clump of vegetation.

This raptor eats mostly mice and voles, but also takes insects and small reptiles. The northern harrier glides close to the ground when hunting, diving down quickly to capture prey.

Northern harriers use their sense of hearing to help locate prey, an unusual behavior among hawks. They have an owllike facial disk to help with directional hearing and soft feathers for a quieter flight.

The call of the northern harrier is a shrill "kek, kek, kek" or "keee, keee, keee."

Relationship to People

Harriers are becoming less common in Narragansett Bay. There are many probable causes including the contamination of eggs by DDT, loss of agricultural and marsh habitat and increased pressure of mammalian predators. The northern harrier is listed as State Endangered in Rhode Island.

FIELD MARKINGS:
Gray upperparts, distinct white rump and white under-parts with spotted breast. Eyes are yellow. Wings are long; gray above and white below with black tips. Size: 16 to 24 inches; wingspan 38 to 48 inches.

HABITAT:
Salt marshes, wet meadows, coastal fields, fresh water marshes.

SEASONAL APPEARANCE:
Year-round.

SENSITIVITY LEVEL:

Osprey

COLLOQUIAL NICKNAMES:
Fish Hawk, Seahawk

SCIENTIFIC NAME:
Pandion haliaetus

Distinguishing Features and Behaviors

The osprey is a large predatory bird common in the coastal areas of Narragansett Bay. Although not a true hawk, the osprey's nickname "fish hawk" comes from a combination of keen eyesight, agility, timing, strong talons and expertise in catching fish. The osprey's talons can turn backward, allowing for stronger grip on prey. Barbs on the soles of the talon help grip slippery fish. Osprey have a slight backward bend in their wings when flying that distinguishes them from other raptor species.

Hovering up to 100 feet above water, the osprey waits for fish to come to the surface. Upon seeing its prey, the osprey plunges down into the water, grasping the fish with its feet. On occasion, the osprey will immerse entirely in the water, a rare behavior for raptors.

Ospreys build large nests of sticks and other matter atop dead trees and poles. Located near or over the water, osprey nesting sites provide protection and help the birds keep a watchful eye for prey. Ospreys mate for life and return to the same nesting site each year. The male and female take turns incubating eggs. Once the young are hatched, the male takes responsibility for providing food.

Relationship to People

Between 1950 and 1970, the chemical pesticide DDT (dichlorodiphenyltrichlorethane) was used on Narragansett Bay salt marshes and wetlands to kill mosquito larvae. Fish fed on the mosquitoes, which had accumulated DDT in their bodies, and ospreys fed on contaminated fish. The pesticide DDT weakened osprey egg shells, causing them to crack during incubation.

Accelerated shoreline development and hunting further reduced the osprey population. Before the 1960s, 100 pairs of osprey were recorded in the Bay. By 1967, there were only four breeding pairs. Since the ban of DDT in 1972, increased environmental protection and the construction of artificial nesting platforms, the population has slowly risen. In 2004, 82 pairs of nesting osprey called Rhode Island home, including the couple seen frequently at Fields Point in Providence.

FIELD MARKINGS:
Brown on top with bright white underside, dark specks on wings and dark bands on the tail feathers. The head is mostly white with a black band extending from the eye through the cheek. Size: 20 to 24 inches long with wingspan of 4 to 6 feet.

HABITAT:
Salt marsh, coastal woodlands.

SEASONAL APPEARANCE:
Early spring, summer, fall.

SENSITIVITY LEVEL:

Belted Kingfisher

COLLOQUIAL NICKNAMES:
Kingfisher

SCIENTIFIC NAME:
Ceryle alcyon

Distinguishing Features and Behaviors

Belted kingfishers are stocky birds with short legs. Their large bills and double-peaked, shaggy-looking crest give their heads an oversized appearance. Across the breast, on both the male and the female, there is a wide blue-gray band for which they get the name "belted." In flight, gray and white bars are visible on the underside of the tail. Belted kingfishers have a loud rattling call.

Primarily fish-eating birds, belted kingfishers also eat aquatic invertebrates, reptiles, amphibians, shellfish, small mammals and young birds as well as insects. When hunting for food, belted kingfishers perch near the water until they spot their prey. They hover over the water with rapid wing beats, then fold their wings back and plunge into the water.

These birds nest in burrows dug horizontally into the side of sandy or muddy banks. Both male and female dig the burrow with their bills and scrape back loose soil with their feet. Sometimes old burrows from previous nests are used again. The female lays her eggs in a chamber that has been excavated at the interior of the burrow. Repeated trips through the burrow entrance give it a distinctive, identifiable shape with two grooves carved at the bottom by adult bird feet. Once they leave the nest, young birds follow their parents around to be fed and to learn how to hunt. Parents teach their young to dive by dropping dead fish into the water where the young can retrieve them.

Except in the breeding season, kingfishers are solitary birds. They maintain a feeding territory for themselves in the fall and winter seasons, driving away all other competing kingfishers, even their own offspring.

Relationship to People

The belted kingfisher is found in many parts of Narragansett Bay. However, because these birds rely on undisturbed sandy banks for nesting, their success is highly dependent upon healthy habitat.

Before laws to protect wild birds were enacted, belted kingfishers were hunted and their eggs collected. Now their population is susceptible to decline due to shoreline development, loss of habitat and water pollution that affects their food supply.

FIELD MARKINGS:
A stocky, blue-gray bird with a double-peaked crest. Broad blue-gray belt on the breast. The female has a rust-colored belly band below the gray belt. Size: 13 inches long.

HABITAT:
Streams, ponds, rivers, coastal inlets, lakes, rocky seacoasts.

SEASONAL APPEARANCE:
Spring through fall; some overwinter.

SENSITIVITY LEVEL:

OCEAN

 ESTUARY

 SHORELINE

WHERE DO I FIND IT?
Use these icons as a quick reference to where you might find a particular Bay species. Icons denote a species' predominant habitat; other preferred habitats are mentioned within the text.

ANATOMY OF A SEAL:

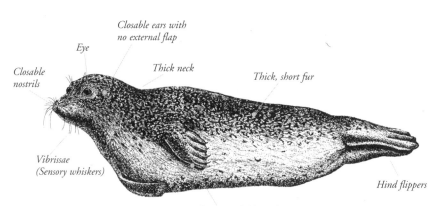

Closable ears with no external flap

Eye

Closable nostrils

Thick neck

Thick, short fur

Vibrissae (Sensory whiskers)

Hind flippers

Fore flippers with blunt claws

Mammals

Several marine mammals are known to visit Narragansett Bay. As with all mammals, marine mammals give birth to live young, nurse their young, have hair, breathe air through their lungs and are warm-blooded. The most common marine mammal in Narragansett Bay is the harbor seal, which visits our area from late fall to spring every year. Here, we highlight the harbor seal along with several other seals that are found seasonally in the Bay.

Seals are very sensitive to disturbances by humans. The smallest, seemingly insignificant disruption can cause them to abruptly leave their haul-out sites and flee into the safety of the water. Disturbance of seals by boaters or onlookers may cause a negative effect on seals' physiology, disrupt the nursing time of their young and cause additional stress in adults. It is a violation of the Marine Mammal Protection Act to inappropriately touch or approach a seal.

Today, all marine mammals in the United States are protected by the Marine Mammal Protection Act of 1972. This act protects these animals from capture or harassment and prohibits the taking of any part of their carcasses. Any action by humans that causes disruption to a marine mammal's normal behavior can be considered harassment and is punishable by fine or imprisonment.

Whales, porpoises and dolphins are marine mammals that prefer large open bodies of water instead of protected bay areas. As an estuary — an ecosystem where fresh and salt water mix — Narragansett Bay is not considered open water. However, it is possible to encounter whales, porpoises or dolphins near the mouth of the Bay and offshore Rhode Island, such as in Block Island Sound.

If you find a seal or other marine mammal that is stranded, dead or being harassed, please contact the Mystic Aquarium at 860-572-5955 or the Rhode Island Department of Environmental Management at 401-222-2284. You will be asked to provide your name and phone number, the exact location of the marine mammal, its species, the size of the mammal and its general condition (injured, disoriented, etc.).

Gray Seal

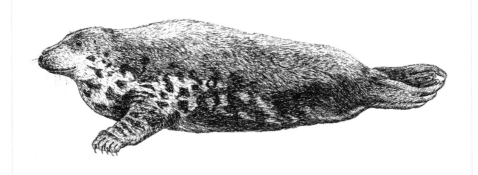

SCIENTIFIC NAME:
Halichoerus grypus

Distinguishing Features and Behaviors

The gray seal's name in Latin translates to "hooked-nose pig of the sea" because of the adult male's long nose and wide nostrils. Gray seals consume a variety of fish, crustaceans and invertebrates, including squid and octopus. Most smaller fish are swallowed whole, whereas larger fish are torn into bite-sized pieces with the claws on the seal's front flippers. Gray seals can swim down to about 475 feet and hold their breath for close to 20 minutes.

Gray seals gather together in large groups and can be observed "hauling-out" on exposed rocks or along the shoreline. During the breeding period, males can be seen fighting with each other for position with the females. During this four-to-six-week period, males and females do not eat. Instead, they live off their fat reserves and concentrate on breeding. Once born, seal pups grow very quickly by drinking their mother's fat-rich milk. Pups are born white with a yellow tint and, on average, they add two to three pounds per day to their body weight.

Relationships to People

Recently there has not been any large-scale commercial hunting of gray seals. However, continued threats to the gray seal population include entanglement in marine debris and the shooting of seals to prevent damage to fishing gear, nets and traps. Nonetheless, population has remained steady, and, in some cases, has increased.

Gray seals are found in open and coastal waters, stretching from eastern Canada to the Baltic Sea, and have been seen as far south as New Jersey. Along with the rise of gray seals in Narragansett Bay and Block Island Sound, there has been an increase in the amount of strandings.

FIELD MARKINGS:
Male coats vary from dark brown or gray to almost black in color, with occasional lighter patches. Females are lighter and tend to be a combination of gray and tan, with occasional darker patches. Size: males can grow up to 10 feet and weigh upwards of 800 pounds. Females can grow to over 7 feet and weigh close to 600 pounds.

HABITAT:
Open ocean waters and coastline.

SEASONAL APPEARANCE:
Late winter to early spring.

SENSITIVITY LEVEL:

Harbor Seal

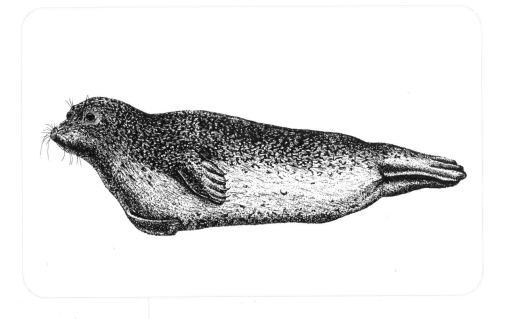

SCIENTIFIC NAME:
Phoca vitulina

Distinguishing Features and Behaviors

The harbor seal belongs to an order of marine mammals called the pinnipeds, which includes seals, sea lions, fur seals and walruses. As mammals, seals are warm-blooded, have hair, and bear and nurse live young. Other marine mammals include dolphins, porpoises, whales, otters and manatees. Seals breathe air but spend a majority of their lives in the water feeding, swimming and mating. Harbor seals inhabit marine, estuarine and freshwater environments. They primarily reside in the colder waters of Maine and Canada, but many travel south during the winter months.

Seals frequently "haul out" onto rocks to bask in the sunlight or rest during low tide. This is the time when they are most often observed. When the tide is high, seals return to the water to feed. They can be very curious and sometimes venture close to boats during foraging. Seals eat mostly fish and squid; their diet varies based on location and seasonal food availability.

When out of water, harbor seals pull themselves forward with the claws on their foreflippers. In the water, they swim using their hind flippers. The profile of a harbor seal is doglike, with a concave forehead and a short muzzle. Compared to other seals with a more parallel structure to the nostrils, the harbor seal's nostrils form a "V" shape, almost meeting at the bottom.

Relationship to People

Seals are not new visitors to Narragansett Bay, but have been seen in greater abundance in recent years. In the late 1800s, seals were hunted for food and for their skin. Additionally, a bounty was offered for the killing of seals, as they were considered a nuisance to fishermen. As a result, the population was almost exterminated by the 1900s.

Harbor seals can be seen at nearly 20 haul-out sites throughout most of the Bay, from Providence and Bristol to North Kingstown and Newport. Curious observers in boats can have harmful effects on harbor seals, often disrupting feeding and resting behaviors.

FIELD MARKINGS:
Light gray or tan fur with black spots and blotches. Size: males 5 to 6 feet and up to 250 pounds; females slightly smaller.

HABITAT:
Open marine and estuarine environ-ments. Found resting on near-shore rocks and islands, in bays, gulfs and other estuaries.

SEASONAL APPEARANCE:
October to May.

SENSITIVITY LEVEL:

Harp Seal

SCIENTIFIC NAME:
Phoca groenlandica

Distinguishing Features and Behaviors

The harp seal's Latin name, *Phoca groenlandica*, means "lover of ice." They can live for up to 35 years and have a doglike muzzle.

Harp seals give birth on pack ice and nurse their young for only eight to twelve days. The mother's milk is about 43% fat. Within a week of birth, a baby harp seal's weight has almost doubled. Pups first molt at three weeks. At four weeks, the pups' coats are silver and flecked with black. The molting process takes about four weeks.

Harp seals are normally found in the North Atlantic, including the waters around eastern Canada and

Greenland. Because they generally do not migrate as far south as Rhode Island, finding a harp seal in the Bay is uncommon. However, in recent years, juvenile harp seals have been sighted in increasing numbers in Narragansett Bay and Block Island Sound.

The majority of harp seals found in Rhode Island waters are malnourished and exhausted from their long migration. They may be found stranded on the Bay's beaches and along the state's southern shore. Scientists do not have definitive answers as to why juvenile seals have been found in increasing numbers so far south.

Relationship to People

Traditionally, harp seals were hunted by native peoples for food, oil and skins. Today, harp seal pups are hunted commercially for their beautiful white coats in Canada. From two and a half to four weeks of age, the pups lose their white hair and their coats are no longer considered commercially desirable. The European Economic Community banned the import of harp seal skins in 1983.

Due to the increase of harp seal strandings in the Bay, there is a greater likelihood of interaction between humans and seals. Do not approach or disturb a seal that is on the shore.

FIELD MARKINGS:
Pups have a white birth coat; juveniles are gray to dark tan with dark spots or markings; adults have a distinct dark harp or horseshoe marking on back Size: pups are 2.5 to 3 feet long, weighing 15 to 25 pounds. Adults are 5 to 6 feet long, weighing 250 to 400 pounds.

HABITAT:
Open ocean waters and coastline.

SEASONAL APPEARANCE:
Fall through early spring.

SENSITIVITY LEVEL:

Hooded Seal

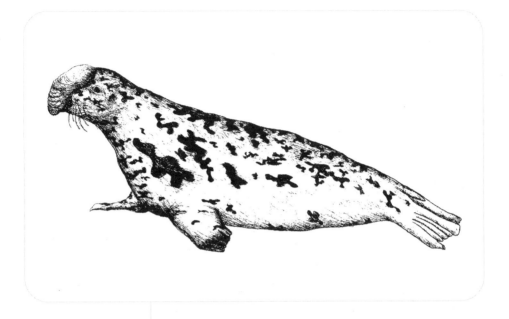

SCIENTIFIC NAME:
Crystophora cristata

Distinguishing Features and Behaviors

The hooded seal's Latin name, *Crystophora cristata,* means "bladdernose" or "crested seal." The male seal's "hood" is actually an enlarged nasal cavity that begins just above and behind the eyes and extends in front of the mouth. When inflated, the hood resembles a leather football. The hood is inflated as part of the courtship display or in response to disturbance.

Hooded seals only nurse their pups for four to seven days, the shortest nursing period of any mammal. The mother's milk is 60% fat and has the lowest percentage of protein of any mammal's milk. It is so rich that the pups gain an average of fifteen and a half pounds in a 24-hour period. Hooded seals mature in three to six years, but may live for 20 to 35 years.

Because they do not normally migrate as far south as Rhode Island, finding a hooded seal in the Bay is unusual. Juvenile hooded seals — usually found in the North Atlantic along the shores of northern Iceland, Labrador and Newfoundland — have been spotted in increasing numbers in Narragansett Bay and Block Island Sound. Having traveled for so many miles, the majority of hooded seals found in Rhode Island waters may be malnourished and exhausted and may strand themselves onshore.

Relationship to Humans

Hooded seals have been hunted since 1874. Before World War II, the seals were hunted for oil and leather. Since then, seal pups have become the primary targets for their beautiful "blue-back" coats. Even though the pups are targeted these days, many adults are killed simultaneously because they aggressively defend their pups. The European Economic Community has banned the import of hooded seal blue-back pelts since 1983.

The likelihood of interactions between humans and hooded seals has increased due to the rising number of hooded seal strandings in Narragansett Bay.

FIELD MARKINGS:
Juveniles are called "blue-backs" due to the steel blue color on the top half of the animal; adults have distinct black patches with black muzzle and face. Size: pups are 3 to 3.5 feet long, weighing 33 to 66 pounds. Adults are 7 to 9 feet long, weighing 660 to 880 pounds.

HABITAT:
Open ocean waters and coastline.

SEASONAL APPEARANCE:
Late fall to early spring.

SENSITIVITY LEVEL:

Whales, Porpoises and Dolphins

BALEEN WHALES

Minke Whale
Balaenoptera acutorostrata

COLOR: Dark gray back with a white underside and gray shading at the sides.
SIZE: 27 to 30 feet long.
SEASONAL APPEARANCE: April to November.
DISTINGUISHING FEATURES: The minke whale has a narrow and "V"-shaped head. It's slim and pointed flippers are marked by a white band.

Fin Whale
Balaenoptera physalus

COLOR: Gray to brown, with white on the underside of its flukes and flippers.
SIZE: 72 to 78 feet long.
SEASONAL APPEARANCE: Summer to early fall.
DISTINGUISHING FEATURES: The fin whale has a large, lighter-colored area behind its right eye. Its back has a distinct ridge from its dorsal fin to its flukes.

Humpback Whale
Megaptera novaengliae

COLOR: Black to dark brown body with a white throat. Part of the belly is also white.
SIZE: 49 to 52 feet long.
SEASONAL APPEARANCE: Summer to early fall.
DISTINGUISHING FEATURES: The portion of the humpback whale's head in front of its blowhole is flat and covered with knobs. It has a fleshy lump on the tip of its lower jaw. The whale has very long flippers, almost one-third of its body length. As its name suggests, the whale's dorsal fin is humped.

TOOTHED WHALES

Harbor Porpoise
Phocoena phocoena

COLOR: White belly with gray sides. The flippers and tail are black.
SIZE: 4.5 to 5.5 feet long.
SEASONAL APPEARANCE: February to October.
DISTINGUISHING FEATURES: The harbor porpoise does not have a beak.
Its flippers are small and blunt, with a triangular dorsal fin that is low and
broad along the base.

Common Dolphin
Delphinus delphis

COLOR: Black back with a white belly. Has a tan and gray criss-cross or
hourglass pattern on its sides.
SIZE: 7.5 to 8 feet long.
SEASONAL APPEARANCE: Year-round.
DISTINGUISHING FEATURES: The common dolphin has a well-defined beak with
a white tip. Its body is very streamlined with a tall and pointed dorsal fin.

Atlantic White-sided Dolphin
Lagenorhynchus acutus

COLOR: Gray body with a white belly and narrow, white patch on its sides.
The dorsal fin is black.
SIZE: 8 to 9 feet long.
SEASONAL APPEARANCE: Year-round.
DISTINGUISHING FEATURES: The Atlantic white-sided dolphin is the most
common dolphin seen off of southern New England. It has curved and pointed
flippers. Its beak is short with a black upper jaw and gray lower jaw.

Glossary

Adductor muscle: a muscle that draws a body part toward the median line

Adipose fin: a soft, fleshy fin located behind the dorsal fin and just forward of the caudal fin in some fish

Agar: a gelatinous material derived from certain marine algae, used as a stabilizer and thickener in many food products; also used in microbiology to grow culture

Aggregate: to collect or gather into a mass or whole

Amphipod: an order of animals that includes over 7,000 described species of small, shrimplike crustaceans

Anadromous: fish that live in the sea but return to fresh water to spawn

Annelid: a large phylum of animals comprising the segmented worms, including earthworms and leeches

Aquaculture: the cultivation of marine organisms in a controlled setting

Arthropod: an invertebrate with jointed legs, a segmented body and an exoskeleton

Audubon Society Blue List: a list published by the National Audubon Society's field journal, American Birds, that provides early warning of the North American bird species undergoing population or range reductions

Barbel: fleshy, sensory projection located near the jaw or mouth of some fish

Benthic: referring to the ocean bottom and organisms that inhabit that area

Bioaccumulation: an increase in the concentration of a particular substance — most often a toxin — in an organism as it consumes those lower on the food chain that have smaller concentrations of the same substance

Bioluminescence: the production of light by living organisms

Bivalve: a mollusk with a two-part, hinged shell, such as a clam, oyste or mussel

Bloom: a dense concentration of phytoplankton that occurs under optimum growth factors, i.e., temperature, sunlight and nutrients

Brackish: water containing a mixture of fresh and salt water; has a lower salt concentration than pure ocean water

Bryozoans: tiny animals that group together in large colonies of many individuals; also known as moss animals or sea mats

Byssal thread: a strong, threadlike fiber produced by mussels to attach themselves to a solid surface or substrate

Calcareous: composed of calcium or magnesium carbonate, which has a rough, stonelike texture

Cannibal: an organism that preys on and consumes members of its own species

Carapace: the portion of an organism's exoskeleton, or shell, that covers the head and thorax region

Carnivore: a meat-eating organism

Carrageenan: a gelatinous carbohydrate used to emulsify dairy products, baked goods and cosmetics

Catadromous: fish that live in fresh water and breed in the sea

Caudal fin: fin located at the tail end of a fish and used for propulsion

Cephalopod: mollusks, such as the octopus or squid, characterized by bilateral body symmetry, a prominent head and a modification of the mollusk foot into the form of arms or tentacles

Chordate: a group of animals including vertebrates, together with several closely related invertebrates; united by having at some time in their life cycle: a notochord, a hollow dorsal nerve cord, pharyngeal slits, an endostyle and a post-anal tail

Chromatophore: a pigment cell that contracts and expands to produce immediate color changes in an organism

Ciguatera poisoning: a form of human poisoning caused by the consumption of subtropical and tropical marine finfish that have accumulated naturally occurring toxins through their diet

Cilia: specialized cells containing short hairlike extensions for locomotion or movement of materials

Clutch: the young of an animal cared for at one time

Cnidocyte: a type of venomous cell unique to the phylum Cnidaria (including corals, sea anemones, hydrae and jellyfish)

Colloblast: one of the cells covered with sticky granules on the tentacles of a ctenophore, which aid in capturing prey

Coenocytic: of a large multinucleate mass enclosed by a single cell wall

Commensal feeding: when one animal helps another find food without wasting any extra energy

Copulation: sexual reproduction between a male and a female organism

Creche: a group of animals providing care to offspring

Crustacean: an organism belonging to the class Arthropoda, which includes crabs and lobsters

Ctenophore: any of various marine invertebrates of the phylum Ctenophora, having transparent, gelatinous bodies bearing eight rows of comblike cilia

Cygnet: a young swan

Dermal denticles: sharp, "V"-shaped scales found on the skin of sharks

Desiccation: the drying out of an organism as a result of moisture deprivation

Detritus: material created from the decomposition of dead organic plant and animal remains

Devonian Period: a geologic period of the Paleozoic era some 400 million years ago

Diatom: microscopic algae encased within a cell wall made of silica

Dioecious: male and female flowers grow on separate plants

Dormant: a state of inactivity that includes a lowered metabolic rate

Dorsal: refers to the back or upper surface of an organism

Dorsal fins: located on the back of a fish and used for balance

Drake: a male duck

Echinoderm: literally "spiny skin;" a class of invertebrates that posess radial symmetry

Eclipse plumage: dull or colorless plumage that certain birds, such as male ducks, acquire at the end of the breeding season

Ecology: the study of the relationship between organisms and their environment

Ecosystem: a community of organisms in their physical environment

Elasmobranchs: an organism belonging to a group of cartilaginous fish consisting of rays and sharks

Elver: a young eel

Epiphyte: one plant living on another plant

Epitoky: a form of reproduction in annelids in which some or all of the segments are turned into reproductive cells and completely split apart

Estuary: the mixing area of salt and fresh water where the river meets the sea

Eutrophic: water enriched with a high quality of nutrients resulting in a high growth of phytoplankton or seaweeds; the decaying of this plant matter sometimes results in the depletion of oxygen from the area, which can be toxic to other organisms

Exoskeleton: a skeleton existing on the outside of an organism, such as the shell of a lobster or crab

Extirpated: to remove or destroy totally; exterminate

Fertilization: the biological joining of a sperm and an egg

Fish ladder: a device constructed to help fish swim upstream to spawn in rivers that have been blocked by dams or other obstructions

Gamete: sex cells

Gastropod: a one-shelled mollusk such as whelks, periwinkles and other snails

Gill cover: a protective plate that lies over the gills of a fish

Gonad: a reproductive organ

Habitat degradation: the overall decline in the quality and/or quantity of a plant or animal's natural home

Hen: a female duck

Herbivore: an organism that feeds exclusively on plants

Hermaphrodite: an animal that carries both male and female sex organs

Holdfast: in plants, a strong rootlike structure that serves as an anchor

Holoplankton: organisms that remain as plankton permanently and do not develop into a different larval form

Hydrodynamic: the ability to propel easily through water without creating much resistance; streamlined

Intertidal zone: the water area between the high and low tide marks

Invertebrate: an animal without a backbone

Iridescent: displaying lustrous colors like the rainbow

Isopod: one of the most diverse orders of crustaceans, most common in shallow marine waters

Larva: the newly hatched, earliest stage of any of various animals that undergo metamorphosis, differing markedly in form and appearance from the adult

Lateral line: a series of sensory detectors arranged along the sides of most fish; detect vibrations and movements of other organisms in the water

Lophophore: ciliated feeding structures, usually round or "U"-shaped, found in some organisms

Lores: in birds, the area between the eye and the base of the bill

Madreporite: the bright orange dot in the center of a sea star, used for pumping water into its body

Marine Mammal Protection Act of 1972: prohibits the harassment, hunting, capture or killing of marine mammals in United States waters and by U.S. citizens on the high seas

Meroplankton: temporary members of the plankton community that are larval forms of organisms

Midden: collection of shell and crab pieces left by octopuses outside their dens

Midrib: a slightly raised, vertical line running through the middle of a blade in some seaweeds

Migratory Bird Treaty Act of 1918: U.S. legislation makes it unlawful to pursue, hunt, take, capture, kill or sell migratory birds; also grants full protection to any bird parts including feathers, eggs and nests.

Mollusk: any of numerous marine invertebrates typically having a soft unsegmented body, a mantle and a protective calcareous shell

Molting: the process of shedding hair, feathers, shell or an exoskeleton periodically

Moratorium: a temporary cessation of an activity, such as fishing, in a particular area

Nektonic: the ability to swim or move independently against currents

Nematocyst: a specialized stinging cell used by an organism to capture prey

Nocturnal: active at night

Notochord: a supporting rod running most of the length of chordates; it is present at varying times in their life cycle

Omnivore: an organism that eats both plants and animals

Operculum: a lid or covering that closes the shell opening in snails, or the bony gill covering of a fish

Ossicles: tiny calcareous skeletal fragments

Overexploitation: the act of harvesting too many fish too quickly for the species to replenish itself

Palp: a small, blunt appendage present in invertebrates for feeding and sensation

Pannes: bare dirt patches within a salt marsh, usually too salty or harsh for most plants to thrive

Parasitism: a symbiotic relationship where the parasite lives in or on the host; beneficial only to the parasite and detrimental to the host

Pectoral fins: pair of fins located near the gill cover; used for maneuvering the fish

Pelagic: the ability to live exclusively in the water column of the open sea or continental-shelf waters, not on the bottom

Pheromone: a chemical substance released by an organism for the purpose of communication between members of the same species

Phytoplankton: planktonic plant species, including diatoms and dinoflagellates; they form the beginning of the food chain for aquatic life

Pincers: a grasping organ or pair of organs resembling this, as the claw of a lobster

Pinnipeds: fin-footed marine mammals, including walruses, seals and sea lions

Piscivore: a carnivorous animal that eats mainly fish

Plankton: free-floating, microscopic sea organisms with limited swimming abilities that float at the mercy of the tides and currents

Pneumatocysts: tiny air bladders in the blades of some seaweeds; they allow the plant to float

Polyandry: a mating pattern where females mate with several males during one breeding season

Predator: a carnivorous animal that feeds by killing other animals

Proboscis: an elongated, extensible tubular structure some organisms use for feeding or sensing

Radula: a rasping, tonguelike organ present in mollusks and used for grazing and breaking up food

Regeneration: the ability of an organism, such as a sea star, to grow a portion of its body back after it has been damaged or severed

Rhizomes: an underground network of stems and roots that anchors a plant and allows it to reproduce by vegetative reproduction

Rookery: a breeding place or colony of gregarious birds or animals, such as penguins and seals

Rostrum: a spiny projection between the eyes of many shrimp

Salinity: a measure of the total amount of dissolved salts in seawater

Scavenger: an animal that feeds on the dead remains of plant and animal matter

Sessile: fixed on or attached to another substrate and, therefore, unable to move

Siltation: the settling of fine mineral particulate matter

Siphon: a tubelike structure present in some mollusks to intake and expel water from the mantle cavity for feeding purposes

Spawning: sexual union between a male and a female for reproductive purposes

Spiracles: a pair of openings on the heads of some sharks and rays used to pump in water that is passed over the gills

Subtidal: the zone present below the low-tide line

Suspension feeder: an organism that uses a filtering system to extract nutrients, such as plankton and bacteria, from the water

Swim bladder: an organ that enables fish to adjust their buoyancy in the water column by filling with gases

Swimmerets: a pair of abdominal appendages of some crustaceans, such as shrimp and lobsters, that functions primarily for carrying the eggs in females and are usually adapted for swimming

Test: the hard, protective shell or covering of certain invertebrates, such as echinoderms or tunicates

Tetrodotoxin: a potent neurotoxin found in some species of fish, including puffers, ocean sunfish and triggerfish

Translucent: allowing light to pass through, but preventing objects on the opposite side from being clearly visible

Thorax: a division of an animal's body that lies between the head and the abdomen

Tunicate: a sessile marine chordate having a saclike body enclosed in a thick membrane, or tunic, with two siphons for the intake and expelling of water

Ventral: refers to the abdominal or underneath surface of an animal

Ventral fins: located forward of the anal fin in fish and used to provide further stability in swimming

Voracious: being indiscriminate in choice of prey and having the ability to consume large quantities

Water column: the water found from the surface to the bottom sediment

Whorl: one of the turns or volutions of a spiral shell

Zooplankton: animal members of the plankton group that float or drift weakly, transported by currents and tidal motion

Index of Common Names

Index of Scientific Names

References

Bavendam, F. 1980, *Beneath Cold Waters: The Marine Life of New England.*
   Maine: Down East Books.

Berry, W.J. 1978. *Aspects of the Growth and Life History of the Sheepshead Minnow,*
   Cyprinodon variegates, from Rhode Island to Florida. University of Rhode Island
   Master's Thesis.

Bigelow, H.B., and W.C. Schroeder. 1953. Fishery Bulletin 74, Vol. 53. *Fishes of the Gulf*
   of Maine. Washington, D.C.: United States Department of the Interior, Fish and
   Wildlife Service.

Cassidy, James, ed. 1990. *Book of North American Birds.* The Reader's Digest Association, Inc.

Chandler, Robert J. 1989. *The Facts on File Field Guide to North American Shorebirds.*
   New York: Facts on File.

Chapman, V.J. 1978. *Coastal Vegetation.* New York: Pergamon Press.

Coulombe, Deborah A. 1992. *The Seaside Naturalist: A Guide to Study at the Seashore.*
   New York: Fireside, Simon and Schuster, Inc.

Donnelly, JP, and MD Bertness. 2001. Rapid shoreward encroachment of salt marsh cordgrass
   in response to accelerated sea-level rise. *Proceedings of the National Academy of*
   Sciences 98:14218-14223.

Ehrlich, Paul R., David S. Dobkin, and Darryl Wheye. 1988. *The Birder's Handbook: A Field*
   Guide to the Natural History of North American Birds. New York: Simon and Schuster,
   Inc.

Enser, Richard W. 1992. *The Atlas of Breeding Birds in Rhode Island.* Providence: Rhode
   Island Department of Environmental Management.

Farrand, John, Jr., ed. 1983. *The Audubon Society Master Guide to Birding.* Vol. 1.
   New York: Alfred A. Knopf.

Farrand, John, Jr., ed. 1983. *The Audubon Society Master Guide to Birding.* Vol. 2.
   New York: Alfred A. Knopf.

Ferren, Richard L. 1995. Personal communication.

Forsyth, Adrian. 1988. *The Nature of Birds.* Camden East, Ontario: Camden House Publishing.

Gosner, K. L. 1978. *A Field Guide to the Atlantic Seashore.* Boston: Houghton Mifflin Company.

Hale, S. O. 1988. *Narragansett Bay: A Friend's Perspective.* Rhode Island Sea Grant.

Hansen, J. 1988. *Seashells in My Pocket.* Boston: Appalachian Mountain Club.

Harrison, Hal H. 1975. *A Field Guide to Birds Nests in the United States East of the Mississippi River.* Boston: Houghton Mifflin Company.

Harrison, Peter. 1983. *Seabirds: An Identification Guide.* Boston: Houghton Mifflin Company.

Hayman, Peter, John Marchant, and Tony Prater. 1986. *Shorebirds: An Identification Guide.* Boston: Houghton Mifflin Company.

Katona, S., D. Richardson, and R. Hazard. 1977. *A Field Guide to the Whales and Seals of the Gulf of Maine.* 2nd Edition. Bar Harbor, Maine: College of the Atlantic.

Kaufman, Kenn. 1990. *A Field Guide to Advanced Birding.* Boston: Houghton Mifflin Company.

Knauss, J.A. 1980. *Maritimes.* Vol. 2. University of Rhode Island Graduate School of Oceanography.

Lippson, R. L., and Alice J. Lippson. 1984. *Life in the Chesapeake Bay.* Baltimore: Johns Hopkins University Press.

Martinez, A. J., and Richard A. Harlow. 1994. *Maine Life of the North Atlantic: Canada to New England.* Massachusetts: Norman Katz.

Matthiessen, G. C. 1995. *Perspective on Finfisheries in Southern New England.* Sussex, Connecticut: The Sounds Conservancy.

Myers, James E. 1995. Personal communication.

Niering, W. A., and R. S. Wareen. 1980. *Salt Marsh Plants of Connecticut.* The Connecticut Arboretum Association.

National Oceanic and Atmospheric Administration (NOAA). 1984. *Distribution and Abundance of Fishes and Invertebrates in Mid-Atlantic Estuaries.* Washington, D.C.: NOAA's Estaurine Living Marine Resources Program, U.S. Department of Commerce, NOAA/NMFS.

National Oceanic and Atmospheric Administration (NOAA). 1994. *Fisheries of the United States, 1993*. Current Fishery Statistics, No. 9300. Washington, D.C.: U.S. Department of Commerce, NOAA/NMFS.

Niesen, T.M. 1982. *Marine Biology Coloring Book*. California: Coloring Concepts.

Olmstead, N. C. 1978. *Plants and Animals of the Estuary*. Bulletin No. 23. Connecticut: Connecticut College Arboretum.

Oviatt, C. 2004. The changing ecology of temperate coastal waters during a warming trend. *Estuaries* 27: 895-904.

Peterson Roger Tory. 1980. *A Field Guide to the Birds*. 4th Edition. Boston: Houghton Mifflin Company.

Peterson Roger Tory. 1980. *Peterson Field Guide to Eastern Birds*. Boston: Houghton Mifflin Company.

Rhode Island National Heritage Program and Rhode Island Endangered Species Program. 1994. *Rare Native Animals of Rhode Island*.

Rhode Island Sea Grant. 1990. *Fish of Narragansett Bay*.

Robins, C. R. 1980. *A List of Common and Scientific Names of Fishes from the United States and Canada*. American Fisheries Society Special Publication No. 12.

Robins, C. R., G .C. Ray, and J. Douglass. 1986. *Peterson Field Guide to Atlantic Coast Fishes*. Boston: Houghton Mifflin Company.

Scott, Shirley L., ed. 1987. *National Geographic Society Field Guide to the Birds of North America*. 2nd Edition. Washington, D.C.: National Geographic Society.

Sheath, R. G., and Marilyn M. Harlin, eds. 1988. *Freshwater and Marine Plants of Rhode Island*. Iowa: Kendall/Hunt.

Simon, A. W. 1987. *Contamination of New England's Fish and Shellfish*. Conservation Law Foundation of New England.

Stokes, Donald W. 1979. *A Guide to Bird Behavior*. Vol. 1. Boston: Little, Brown and Company.

Suprock, Lori. 1995. *Northeast Nesting Ospreys*. Osprey Newsletter 51.

Sumich, J. L. 1988. *An Introduction to the Biology of Marine Life*. Iowa: Wm. C. Brown Publishers.

Tedone, D. 1981. *The Complete Shellfisherman's Guide: Maine to Chesapeake Bay*. Connecticut: Peregrine Press, Publishers.

Terres, John K. 1980. *The Audubon Society Encyclopedia of North American Birds*. New York: Wings Books.

Twining, James E. 1987. *Mute Swans of the Atlantic Coast*. Wickford, R.I.: Dutch Island Press.

Veit, Richard R., and Wayne R. Petersen. 1993. *Birds of Massachusetts*. Massachusetts Audubon Society.

Villalard-Bohnsack, M. 1995. *Illustrated Key to the Seaweeds of New England*. Kingston, R.I.: Rhode Island Natural History Survey.

Weiss, H. M. 1995. *Marine Animals of Southern New England and New York*. Connecticut: State Geological and Natural History Survey of Connecticut.

White, C. P. 1989. *Chesapeake Bay Nature of the Estuary: A Field Guide*. Maryland: Tidewater Publishers.

White, L. B. 1960. *Life in the Shifting Dunes: A Popular Field Guide to the Natural History of Castle Neck, Ipswich, Massachusetts*. Boston: Museum of Science.